In The Beginnings

The Story of the Original Earth, its Destruction, and its Restoration

A Defense of the Biblical Gap Theory of Creation

Steven E. Dill, D.V.M.

Xulon
PRESS

Scripture References

INTRODUCTION

—∿∿—

This book is about Earth's two beginnings. Putting it as bluntly as I can, it is a defense of the Gap Theory of Creation. This theory is also called the Ruin-Restoration Theory, or simply the Restoration Theory, because it deals with the destruction of the original Earth and its later restoration. It is called the Gap Theory because it holds that there was a gap of time between the earth's original creation and its later restoration. This theory is very unpopular these days. I say this right at the start because I don't want you to waste your time reading it if you have already decided you hate the Gap Theory. Christians have become extremely divided over the "correct" interpretation of God's Word concerning creation, and it has become a very emotional topic. Christian Creationists have divided themselves into camps huddled around various creation theories. These camps are so divided that people often want to know your stand on these theories before they're willing to accept you as a fellow creationist, or sometimes, even as a fellow Christian. Many good Christians have been labeled as evil people simply because they hold opinions about creation that differ from someone else's. I suspect I will be one of them.

Ironically, this book had two beginnings. Its first beginning was as a final chapter in another book, *Objections to the Doctrine of Evolution, Book Two*.[1] That book was written to show how Scripture disproves the Theory of Evolution. It's predecessor, *Objections to the Doctrine of Evolution, Book One* was written to show how science disproves the Theory of Evolution. These books are an apologetic for Biblical creation. At first, I didn't mention the different creationists' theories. They seemed relatively unimportant compared to the issue of exposing evolution as a fraudulent theory. Eventually I added a chapter on the Gap Theory because people wanted to know my opinion. I knew more than a few creationists would reject it, but I thought it might help unbelievers. **Much of the unbelievers' objection to biblical creationism is based on the fact that some creation theories are not only unscientific, but anti-scientific.** I wanted unbelievers to know there was at least one creationist theory that agreed with true scientific observations. It wasn't long before I realized that adding it to the other books was a mistake. It made the books too long. A section defending the Gap Theory made the books unpublishable. It needed to be its own book. Although I decided to give my defense of the Gap Theory a new beginning, some of the material in this book remains in the other books as well. Some arguments are useful in both places. For example, I show how the six days of creation in Genesis are literal, twenty-four hour days. This is useful both as biblical proof against evolution and as a defense of the Gap Theory against the Day-Age Theory.

The purpose of this book is simple. I want you to know the truth about our Creator. The Word of God reveals the truth about our Creator, but so do the Works of God. I believe God is smart enough and powerful enough and sovereign enough to make His Works and His Words agree. If you read the comments from the various Creationists' Camps, you get a feeling that this isn't true. There is so little agreement among creationists about science and Scripture that it makes you wonder if God didn't make a mistake when He created creationists. I believe all the bickering harms the Gospel of Jesus Christ more than it helps. I believe it ought to stop. I believe that only truth can make it stop. I believe the Gap Theory is true. It would be nice if I could convince all the other creationists to agree with me about the Gap Theory, but that's not my goal. My real goal is to help unbelievers learn about our Creator.

My Thanks

Above all others, I thank my Lord and Savior Jesus Christ for bearing my sins on the cross and for giving me His righteousness and eternal life. I also want to thank Him for sending the Holy Spirit to live in me and for His guidance and revelation in spiritual matters. Without that gift, I could not understand even the most simple spiritual concept. I give God the credit for all that I am and all I have written.

I also want to thank my good friend Kris Berry of Louisville, KY for reading the book and pointing out a number of inconsistencies and mistakes. He also provided valuable insight into how I should express certain ideas and thoughts. Thank you, Kris for your help.

This book would not have been possible without the aid, understanding, and encouragement of my wife, Linda. She was a valuable proofreader of the manuscript, ever looking for errors and ways to improve the finished product. More than that, she is a valuable proofreader of my life and my Christian conduct, helping to improve that finished product as well. Thank you, Linda, for always being there and for the sacrifices of time, money, and energy that it took to get this book published. Thank you, Jesus, for giving me this wonderful woman to be my life partner. Without her, I could not have written this book.

I wish to thank the people at Xulon Press. They all cheerfully went the extra mile in helping me publish my book. I encourage anyone who wants to be a published author to call on Xulon. You won't be disappointed. —Steven Dill

Table of Contents

—ᵐᵐ—

Chapter One:

The Truth About Truth

—∿∿—

What really happened in the beginning? What did God do when He created things? What does the Bible say about creation? Those are big questions, and frankly I don't have all the answers! I know some of the details, but not all. Only God knows it all. I know creation wasn't by evolution. I know God created various kinds of living things, and I know they have remained reproductively faithful within the limits He set. I know Adam and Eve were real, and I know they were the parents of us all. Biblical Creationists all over the world would agree with these things, but there are disagreements on the details.

One point of disagreement for instance is what the word "kind" means in the Bible. God created life forms to multiply "after their kind." Since God created plants and animals to reproduce in "kind," some believe that all the members of a "kind" should be cross-fertile with all the other members of the same "kind." Some assume God created only one species in each "kind." They think the different species seen in any particular "kind" today are the results of subsequent mutations and natural selection. For example, they assume that all the various species of owls are descendants of one original species of owl. Likewise, all the various butterflies have descended from one original species of butterfly. But, if you assume God created only one species per "kind," you soon find yourself in an awful plight trying to explain speciation, the development of new species. You find yourself accepting the vague concept of micro-evolution. I've heard a number of creation-scientists admit that micro-evolution is true. Evolutionists love such concessions. To them, the difference between micro-evolution and macro-evolution is only a matter of degree. Theistic evolutionists point to the tremendous variety of species and subspecies in the taxonomic groups, and then use this as evidence for evolution. After all, if God started with only one (or even a few) species per "kind," and today there are numerous species per "kind," then there had to be some evolutionary (change over time) process down through the ages. Many creationists have had a hard time trying to explain speciation. Now, I'm not saying there isn't such a thing as speciation. There most certainly is, but I'd be careful about using the terms "micro-evolution" and "macro-evolution." Micro-evolution and macro-evolution have a wide range of definitions depending on who uses them and why. What one evolutionist means by "micro" may be the very same thing another evolutionist means by "macro." Whether speciation is micro-evolution or macro-evolution is unimportant. Speciation is not proof for evolution unless you first prove that evolution and only evolution could create the species seen within the "kinds." You would have to prove that a Creator couldn't have done it. In other words, you need to prove that evolution is true before you could use speciation as proof for evolution. Furthermore, speciation is not evidence that the Bible is wrong because the Bible doesn't define "kind" in scientific terms. We don't know the actual limits God placed on reproductive variation. The Bible doesn't say what "kind" means? It could have a number of meanings. In some cases it may refer to species. In other cases it may refer to genus. It may even refer to family or order sometimes. I don't think we should assume that "kind" refers only to species. I don't think we should try to limit God's creative abilities. Neither should we question God's choice of the word "kind" simply because it doesn't

describe the Linnaean system of taxonomy. This may seem
like a ridiculous point to worry about, but believe me, there
are many creationists who get hung up on what "kind" is
supposed to mean. I don't know what life forms He created,
but none of them were the products of random mutations and
natural selection.

There are a lot of disagreements about the creation, but
the biggest disagreement involves the time factor. When did
God create? How old is the universe? I want to share my
opinion with you, but I also want to share my motive. You
see, while it is important to know WHAT someone believes,
sometimes it's more important to know WHY he believes
it. When you learn why someone believes something, you
often learn why he accepts or rejects new information. Few
are willing to let new evidence change their preconceived
ideas. Why am I saying this? There are two reasons. First,
because over the years I have learned that when it comes
to believing something, motive trumps facts almost every
time. Television celebrity Art Linkletter once hosted a television program that included a segment called, "Kids Say the
Darnedest Things." It was a humorous look at how young
children expressed their thoughts about the world around
them. Equally funny would have been a segment called,
"Adults Believe the Darnedest Things." People will believe
and defend some of the most ridiculous ideas as long as
they are motivated to do so. I have named this phenomenon
"Mental Inertia." Its definition is: An idea believed tends to
stay believed, and an idea rejected tends to stay rejected.
Its corollary is: It takes less evidence to convince someone
they are right than to convince them they are wrong. The
clearest, plainest, most-obvious fact can be ignored and
rejected by the slimmest of evidence if that fact threatens
a cherished opinion. My hope is that you will always check
your motive when faced with any evidence in this book that
is contrary to what you believe. The second reason I mention

this is because I haven't always believed what I now believe. At one time I was a theistic-evolutionist. My motive was because I believed in evolution. God needed billions of years to evolve life on earth. Then I learned a remarkable truth. The Theory of Evolution was a lie. Scripture and science disprove it. I then became a Young-Earth Creationist. My motive was because I assumed the Bible taught that the earth was young. Later, as I probed deeper into Scripture, I became less sure of earth's age. I began to realize how the Bible might reveal that the earth was much older. What I used to know I knew, I now know I didn't know. After much study, I realized the Gap Theory made the most sense. (If you're not familiar with the Gap Theory, I'll explain it shortly.) The Gap Theory resolved a lot of disagreements, and it didn't disagree with true biblical revelation or true scientific observations. Well, let me clarify that statement. The Gap Theory I eventually accepted doesn't disagree with science or the Bible. Like most creation theories, the Gap Theory has numerous sub-theories and variations. Some of these sub-theories truly disagree with science and Scripture. So, I will not try to defend all forms of the Gap Theory. I will defend only the Gap Theory that agrees with the evidence of Scripture and the evidence of science. That evidence changed my preconceived ideas, and if you disagree with the Gap Theory, I hope it will change yours as well. If our goal is to find truth about the creation, then we need to examine all the evidence. We need to examine both scientific and Scriptural evidence, and we need to study how the different schools of thought explain that evidence.

Before we do that, I want to emphasize that I am not going to drag in the arguments of evolutionists, those who don't believe the Bible, and those who simply want to argue. I will deal only with opinions held by creationists who are believers in the Lord Jesus Christ. I am going to limit this to Christian Creationist theories. Now, I may not have reached the same conclusions as some of my brothers and sisters in Christ, and my theory may not agree with their theories, but that doesn't mean they aren't Christians. They are just as saved and just as loved by God as I am. If you are a believer in The Lord Jesus Christ and you think that Genesis is best interpreted by saying, "God laid an egg and the universe hatched out," then I praise God for you and your opinion. Furthermore, I pray you can use such an interpretation to the glory of Jesus Christ. However, if you belong to some church, denomination, or organization that makes defending your Hatched-Egg Theory more important than defending the integrity of biblical revelation, then I have a warning for you. Be careful you don't elevate your love for scholarly recognition above your love for truth.

I want to start by discussing the nature of divine revelation. What does the Bible reveal about creation? There are many opinions. I have one opinion. You may have another. What is the truth about creation? Well, a funny thing about knowing the truth is that you have to know the truth to know whether it's true or not. For instance, if I told you I was a veterinarian, you really wouldn't know it was true unless you first knew whether or not I was a veterinarian. If you knew I was a carpenter, then you'd know my claim of being a veterinarian was a lie. If you knew I really was a veterinarian, then you'd know my claim was true. You wouldn't know the truth about my claim unless you first knew what was true. However, if you knew I was an absolute truth-teller who would never lie, then you would know I was a veterinarian simply because I revealed it to you. You wouldn't need to know my occupation in order to know my occupation. Instead of knowing something about my occupation, you'd have to know something about me. You'd have to know I was an absolute truth-teller. (Which by the way, I am not, and that's the absolute truth.) Now, don't get me wrong! I'm not trying to be relativistic and say there's no such thing as truth. Quite the contrary, there is such a thing as real, objective, absolute, universal, timeless truth, and that's the problem. The only way we can know what is true about creation is by knowing what is revealed by the only One who knows and tells the truth. We can't know it by ourselves. Only God knows the truth about creation, and only by divine revelation is it possible for any of us to know even the tiniest portion of the truth. However, it is the meaning of that very same divine revelation that causes our disagreements. This is our dilemma: either God is a poor revealer of truth or we are poor understanders of truth.

I hope you didn't have to ponder very long on which it is. There's nothing wrong with God. God revealed the truth to us quite well, thank you! Oh, He didn't intend to reveal every facet and detail about creation. He chose to reveal only a tiny fraction; that's all our little brains could understand. What He has revealed, however, is true, and is knowable to the limits He chooses to let us know. Beyond that, we can't know the truth. Some don't like this. Some don't like the way God reveals information to us. Believe it or not, I've read the comments of more than one creationist expressing dissatisfaction with the way God revealed information about the creation in the Bible. They think the creation accounts don't reveal enough technical information, or that they're too simplistic. Phooey! I see nothing wrong with a God who intended to reveal information in a non-technical way. Not only is it not incorrect, but it is quite clever to be able to reveal information about the creation in a way that both modern man and primitive man could understand. I see nothing wrong with a perfect God who intended to reveal only a portion of the truth. I see nothing wrong with a perfect God who says, "Okay, I'll let you know this much right now, but beyond that, you'll have to wait until you are able to understand more." I see nothing wrong with a perfect God who gives a perfect revelation of truth yet prevents it from being fully understood until the time He sees appropriate. After all, this is what He did with the Messianic prophecies. Christ was perfectly revealed in the Old Testament, but no one, not even the prophets themselves, understood all that had been revealed. Even John the Baptist didn't know what had been revealed. John didn't fully understand Jesus until after Jesus revealed to him what He had fulfilled. The prophecies came first, Christ fulfilled the prophecies next, and understanding the prophecies came last. **It wasn't until Christ fulfilled the prophecies that the prophecies about Christ were understood.** Divine revelation often has the mysterious quality of being fully revealed but only partially understood until such time as the Holy Spirit chooses to let it be known. What's the meaning of "six hundred sixty six" in the Book of Revelation? There are many opinions, but no one really knows. Once the times are fulfilled, however, its true meaning will be quite evident.

It seems that God often holds back the understanding of revelation until more information has been revealed. The Jews were looking forward to Messiah coming as the Conquering King, but not to His coming as the Suffering Savior. His saving ministry was fully revealed in the Old Testament, but the revelation of the Saving Messiah wasn't understood until after the Saving Messiah was revealed. It wasn't until after God came as the Savior that the saving ministry of God was comprehended. God had repeatedly told them that He alone would be their Savior.

Isa. 43:11 "I, *even* I, *am* the LORD; and beside me *there is* no saviour." (KJV)

The Jews who believed in Jesus understood this. They understood that if God was their Savior and if Messiah was their Savior, then Messiah had to be God. They understood that Jesus was both LORD God and Messiah. That's why they accepted Him as their King. This understanding escaped the Jews who didn't believe in Jesus. They rejected what God revealed in their own Sacred Scriptures. They didn't want a Suffering Messiah for King. Rather than accept The Son of David as King, they cried out, "We have no king but Caesar." (John 19:15) In doing this, however, they inadvertently exposed their rejection of God's Written Word. God had told them through Moses that only an Israelite could be their king.

Deu. 17:15 "You shall surely set a king over you whom the LORD your God chooses, *one* from among yourselves; you may not put a foreigner over yourselves who is not your countryman." (NASB)

They rebelled against God's revelation because it was contrary to what they wanted to believe. Instead of believing what the Bible ACTUALLY said, they believed what they WANTED the Bible to say. They accepted only the truth that met their criteria. It had to agree with what they wanted to believe before they would believe it; Mental Inertia. The Bible reveals not just what they believed, but why they believed it. The Jewish leaders who rejected Christ were men of power and prestige in their culture. They were the educated elite. They were honored and respected by the people, and they didn't want to lose that standing. They couldn't acknowledge that some uneducated carpenter's son from Galilee could be Messiah the Prince. They refused to believe the Written Word had revealed such a thing. Why did they reject Jesus? If they admitted that Jesus was Messiah, then they would be admitting that they were rejecting the Written Word. This was their motive. Their faith was not in God's Written Word but in their own interpretation of God's Written Word. They ignored the fact that God was capable of revealing truth in ways other than how they thought truth should be revealed. God's revelation took on a form other than a Written Word.

God's revelation became a Living Word. All they had to do was look at how Christ lived. His life revealed His Deity. When John the Baptist was in prison, he sent his disciples to question Jesus about His identity. John knew the written revelation, but he still wasn't absolutely sure who Jesus was. In response, Jesus didn't send John's disciples back to him with a theological treatise. Instead, He sent them back to tell John what they saw Him do. (Luke 7:19-23) His Deity was revealed in what they observed. They already knew the prophecies, but they didn't understand them until they saw Jesus fulfilling them. The unbelieving Jews failed to learn the truth about their Messiah because they ignored what they saw with their eyes. They clung to their preconceived ideas of what Messiah was supposed to be. He was supposed to be the Conquering King, and yes, someday He will come as the Conquering King, but He first wanted to come as the Suffering Savior. God gave them eyes to see this, but they refused to use them. They chose to accept only the prophecies of Messiah the Conquering King. They chose to reject and ignore the prophecies revealing Him as a Suffering Savior. They chose to believe their own preconceived ideas of how Messiah should be revealed rather than how God chose to reveal Him. God gave them a chance to learn the truth by using their eyes, but they shut their eyes to the truth. They held to their presuppositions and failed to see what God revealed. They failed to understand Messiah because they failed to understand the revelations (plural) of Messiah. Our problem with understanding the creation is similar. We don't understand the creation because we fail to understand the revelations (plural) of the creation. We ignore the same two things about God that the unbelieving Jews ignored.

First, we ignore the fact that God reveals truth in various portions of Holy Scripture. All the truth about creation is not in Genesis alone. If I told you I worked with animals, you wouldn't know I was a veterinarian. I might be a marine biologist, a zoo keeper, or a cowboy. I would have to reveal more about my profession before you knew what I did. If I wrote you a series of letters, and each letter contained a clue to my occupation, then you'd have to assemble the letters to get a better idea of my job. You'd be foolish to read only one letter and decide what I did for a living. The same is true of creation. We need to see what the entire Bible says about it. Some people think they can take one or two sentences out of Genesis and fully comprehend the vastness of our Creator's actions and purposes. How foolish! We must look at what the whole Bible reveals. We must let it speak for itself, and we must accept everything it teaches, not just those parts that defend our preconceived ideas.

The second thing we ignore about God is that He has given us eyes to observe His handiwork and brains to understand it. Genesis 1:1 revealed the truth about the space-time continuum, (See *Objections to the Doctrine of Evolution, Book Two.*) but who understood it until Albert Einstein was able to deliver that knowledge to us? Einstein and other physicists made the space-time continuum a scientifically observable phenomenon, but it had been true all along. God created the space-time continuum at the beginning. Space-time physics was just as true in the time of Adam as it is today. However, God's revelation of space-time physics wasn't understood because human eyes had never observed it. The space-time continuum was an unknown phenomenon of God's creation even though He had revealed it through Moses. Oh, a few of the great historical theologians like Nahmanides (A.D. 1194-1270, Jewish philosopher, scholar, and theologian, born in Spain) wrote about it, but they were ignored. Only after science observed it, did we realize the Bible had revealed it from the very first sentence in Genesis. The first sentence in the Book of Genesis is not a simplistic statement made by primitive nomads. It is as profound and information laden as Einstein's formula $e=mc^2$. It is a very well worded message that on one level reveals historic information for primitive man, but on another level reveals scientific information for modern man. This additional scientific revelation had to wait until scientific man used his eyes and his brain to observe the universe. Ironically, the space-time continuum was one of the first things God revealed to man but one of the last things understood. It made sense once we understood the space-time continuum. There could have been no such thing as space or time before creation. Everything that exists was made by God. Nothing but God existed prior to creation. God is spirit, not space; God is eternal, not time. God has no time or space. There would have been no space or time until Jesus brought them both into existence as the space-time continuum. This insight gives us a better understanding of the universe. It is constrained by the confines of time and space. Genesis 1:1 also helps us understand our Creator. He is not constrained by space or time. His ways are not our ways.

The Bible revealed other clues about our universe such as the relationship between matter and energy, the force of gravity, and the principles of thermodynamics. None of these things were appreciated until scientists came along and gave us a greater understanding of nature. Many creationists like to use the Laws of Thermodynamics to prove that evolution couldn't have happened. I am one of them. However, I haven't forgotten that it was scientists and not theologians who discovered the Laws of Thermodynamics. The principles of thermodynamics had been revealed in the Bible for centuries, but those portions of Scripture weren't understood by theologians until after scientists discovered the principles of thermodynamics. We had to observe our universe to see how thermodynamics worked before we realized that the Bible had already told us how thermodynamics worked. The Bible had given us clues to these things, but none of these things were understood until after people used their eyes to observe them. Careful, precise observations have added to our understanding of God's creation. The heavens do indeed reveal His handiwork.

Because of this, I think it is wrong to discount scientists and their observations if we want to know more about our universe and its beginning. We cannot rely on Bible scholars alone. No matter how well and how carefully the Hebrew and Greek exegetes ply their craft, they cannot reveal all that can be known about the creation. Everything there is to know about the creation is not found in the Bible. The Bible doesn't mention neutrinos, bosons, or anti-matter, but God still created them. God didn't choose to reveal quantum physics, black-holes, fundamental forces, quark dynamics, or string-theory in the Bible. However, He did reveal these things by giving us eyes and ears and brains enough to discover them as we studied His handiwork. We have learned things about God's creation by careful interpretation of His Word, and we have learned things about God's creation by careful observation of His Work. Naturally, we must hold that the Bible and science harmonize, but we must be very cautious. When I say the Bible and science harmonize, I mean what the Bible truly says (not necessarily what we interpret it to say) and what science truly reveals (not just how we interpret observations) will not be contradictory. They may be complementary, but not contradictory. The heavens declare the glory of God, but we have to be careful in our observations. All Scriptures are God-breathed, but we have to be careful in our interpretations. We need to look at both revelations of God's creation. We need to look at what God reveals about the creation from the Bible. We also need to look at what God reveals about creation from scientific observations.

What do scientific observations reveal about the creation? We need to see what scientists have observed. More specifically, we need to see what creation-scientists, Christian men and women who are true creationists and true scientists, say they have observed. Since we're dealing with such a small group of people who all swear they're trying to glorify Christ, you'd think there wouldn't be much controversy. Alas, it is not so. Battle lines have been drawn and war wages between two major camps of creation-science. These are the Young-Earth and the Day-Age camps. There are other views of creation, but scientific creationists generally belong to one of these two camps. The Young-Earthers believe God created the universe somewhere in the range of 6,000-8,000 years ago. Young-Earthers have a wide variety of dating techniques that seem to prove the earth is quite young. The Day-Agers believe creation happened about 13 billion years ago with the earth being formed about 4.5 billion years ago. Day-Agers have a wide variety of dating techniques that seem to prove the earth is quite old. I have read the books and heard the arguments from both sides. Both camps have some strong evidence for their beliefs. Both camps have some pretty flimsy arguments as well. So how do we determine what's true?

The best way to find truth is to approach the Bible and science from as unbiased a position as possible. This doesn't guarantee we'll make correct conclusions, but it helps eliminate blind spots in our thinking. Often, we overlook an opponent's arguments simply because we already "know" he's wrong. I shouldn't presuppose that one particular interpretation of science is right or wrong until I see what the Bible truly reveals. Likewise, I shouldn't presuppose that one particular interpretation of the Bible is right or wrong until I see what science truly reveals. Evolution is wrong, but that doesn't mean science is wrong. We must not automatically reject what scientists tell us about the age of the universe just because it doesn't fit our interpretation of Scripture. We can reject it if it contradicts biblical truth, not just biblical interpretation. The best way to discover the truth about the creation is to make absolutely certain our biblical interpretations are correct and our scientific observations are valid.

I am a creation-scientist. This means I believe we should sift our opinions of **science** through the filter of true biblical revelation. If there are a dozen scientific theories explaining how the universe got here, we can automatically exclude those that are truly contrary to truly revealed biblical facts.

I am a creation-scientist. This means I believe we should sift our opinions of **the Bible** through the filter of true scientific observation. If there are a dozen theological theories explaining how the universe got here, we can automatically exclude those that are truly contrary to truly observed scientific facts.

Please note: I do not believe that human observation and Divine revelation are equal in authority. Much has been said about how the domain of religion and the domain of science reveal different aspects of creation. (Science says "how" and the Bible says "why.") Much has been said about how science and religion should stay in their respective domains and not overlap. Much of what has been said is pure human foolishness. Human observation is limited, but Divine revelation is only as limited as God decides it to be. When God reveals something to us, we can be sure it is true whether or not we can verify it by repeated scientific observations. Miracles are true even if they seem to contradict all previous human experiences. Natural observation reveals that God can create a planet with grapes growing on it, and that these grapes can be used to make wine. Miraculous observation reveals that God can make wine without grapes. When Jesus made wine directly from water, it wasn't to teach us that our normal observations are wrong. It was to teach us that our normal observations are limited. The Bible clearly reveals that God is not a God of deception. Generally, we can trust our observations, but not always our interpretations. The only reason I can have faith in true scientific observations is because God has declared that observation is a valid system of perception. The Bible tells us we can learn things about God because He has made Himself known in His works. He is revealed by the works of His hands.

Psa. 19:1-3 "For the director of music. A psalm of David. The heavens declare the glory of God; the skies proclaim the work of his hands. {2} Day after day they pour forth speech; night after night they display knowledge. {3} There is no speech or language where their voice is not heard." (NIV)

Rom. 1:20 "For since the creation of the world His invisible attributes, His eternal power and divine nature, have been clearly seen, being understood through what has been made, so that they are without excuse." (NASB)

Physical observations can reveal truth provided we don't superimpose our philosophies and beliefs on what we observe. "He who has eyes, let him see; and he who has ears, let him hear," are not wasted words. What the eyes see and the ears hear, the heart can twist. We're all good at doing this. My faith in observation isn't founded on the assumption that observations are inherently reliable. My faith in observation is founded on the belief that God hasn't created an observable universe merely to trick us, confuse us, and lead us into ignorance. When I see a rock, I can be certain the rock exists because I have first assumed that God isn't a God of confusion. When scientists make true scientific observations, I can be certain what they observe is true because of that same basic assumption. If you are a creationist, then I implore you not to reject what scientists discover when they observe the universe. God just might be revealing His invisible attributes. I remember reading the comments of some creationists who were overjoyed when the Hubble Space Telescope was discovered to have had a defective lens. They boldly claimed it was God's will that the Hubble Project fail. In truth, they were afraid of what scientists might learn about the universe. They were afraid of what the universe might reveal about its origin. Can you believe that? They were afraid of what God's handiwork might display. How foolish! The things we have learned from the Hubble telescope, once its lens was corrected, have been a tremendous apologetic for the Bible. I would ask my readers to review their motives very carefully before they reject what scientists have truly observed in our universe that might indicate its age. What science has revealed about the age of the universe may provide the key needed to unlock the truth about creation.

That, my friends, is all I'm going to say about the age of the universe for now. I'll reopen that subject later, but if you want more details, go to your local Christian bookstore. You will find plenty of books about the age of the universe. Be warned! Opinions vary tremendously. There are a great number of books defending the views of the Day-Age gang, and there are a great number of books defending the views of the Young-Earth gang. There are also a few others that don't quite fit either view. Some of them are quite interesting.

Chapter Two:

The Creation Combo Special

—∽∿∽—

I would like to present some of the different beliefs about the biblical account of creation, but I don't want to get tangled up in all the theories, sub-theories, and variables that have been fought over in the past. There's no future in that. It would be too complex and it would make this book so long that even I wouldn't read it. So, instead of trying to examine every possible variable, let's look at four questions concerning the creation account. It is the disagreement over these four major characteristics of creation that has led to the various creation theories. If we can determine how the Bible truly answers these four questions, then we will be much more likely to understand what the Bible truly reveals about the creation. Here are the four questions:

1. What does "day" mean?
2. What does "create" mean?
3. What does "beginning" mean?
4. What does Genesis 1:1 mean? (In relationship to the rest of the chapter)

Each of these four questions involves different aspects of creation. Each has multiple variables. Note that the variables aren't necessarily mutually exclusive. Pick one variable from each of these four characteristics and you can create your own creation theory. It would be too confusing and time consuming to sort through all the combinations.

Instead, let's look at how some of the more prominent theories fit together. Here are the variables in outline form to keep things organized. A more detailed explanation of these variables follows the outline. Refer back to this outline if things get confusing. Hopefully, it will help keep it all in perspective.

I. THE MEANING OF "DAY"
    A. Day-Age (Requires an Old Universe)
        1. Gradual-Creation (Progressive Creation Theory)
        2. Periodic-Creation (Punctuated Creation Theory)
    B. Twenty-Four Hour Day (Either an Old Universe or a Young Universe)
        1. Old-Universe
            a. Dual Creation (Gap Theory)
            b. Punctuated-Creation (Multiple-Gaps Theory)
        2. Young-Universe, Single-Creation (Young-Earth Theory)
    C. Days of Relativity (Age of the universe not specified; usually used to defend an Old Universe)
    D. Days of Revelation (Age of the universe not specified; usually used to defend an Old Universe)
    E. Days of Divine Decrees (Age of the universe not specified; usually used to defend an Old Universe)
    F. Repeated-Days (Age of the universe not specified)

II. THE MEANING OF "CREATE"
    A. Co-Eternal Matter and Energy
    B. *Ex Nihilo* Creation

III. THE MEANING OF "BEGINNING"
    A. Long Period of Time
    B. Instantaneous Point of Time

IV. THE MEANING OF GENESIS 1:1 (WHEN COMPARED TO GENESIS 1:3-31)
   A. Title
   B. Summary
   C. Creative Act

I. THE MEANING OF DAY

A1. Day-Age, Old-Universe, Gradual-Creation (Progressive-Creation Theory)

According to this sub-theory, Genesis describes God's creation of the universe, the earth, and life on earth. Creation is a very slow process taking place over several billion years. The Big Bang Theory fits nicely with this theory. The creation is divided into smaller periods called "days" that are themselves millions or billions of years long. The "days" represent general types or categories of God's creative acts. During the course of the "days" God continually created, but did not evolve, the changing life forms that have lived on this planet. He continually and gradually changed the species living on the earth. He gradually created new species and gradually destroyed old species. According to this theory, the geological strata were gradually formed during these long day-ages. This theory is popular and is defended by a large number of good Christians. While it is a creation theory, it can be so close to theistic evolution that you might confuse the two. In fact, if the continual creation events were gradual enough, it would be impossible to distinguish continual creation from theistic evolution. (Theistic evolution is the theory that life on earth actually evolved by natural forces, but that it was planned by God.) I don't accept the Day-Age Theory because I am fully convinced that these days were twenty-four hour days. In addition, this form of the Day-Age Theory has three problems with geology. I'll explain these problems in the next chapter, but for now, here they are:

1) The order in which fossils appear in Geology does not match the order of creation listed in Genesis.
2) It doesn't explain gaps in the fossil record.
3) It doesn't explain living fossils.

A2. Day-Age, Old-Universe, Periodic-Creation (Punctuated-Creation Theory)

This is similar to the above theory. The universe is billions of years old, but in this case, God made it in bursts of non-continual creative acts. God periodically came to earth to create new life forms, and possibly to destroy older ones. Again, the six "days" of creation are six long ages representing six general types or categories of God's creative works. As a result, life was directly created in a punctuated fashion, not gradually, and not by evolution. This theory is the creationists' equivalent of the Punctuated Equilibria Theory of Evolution. (I'll explain Punctuated Equilibria later.) In fact, it becomes difficult to distinguish between Punctuated Evolution and Punctuated Creation. This theory is believed by a lot of good Christians too. On the positive side, it explains the gaps in the fossil record, but it suffers from the other two geological problems. The "days" are still in the wrong order and it can't explain living fossils.

B1a. Twenty-Four Hour Day, Old-Universe, Dual-Creation Theory (Gap Theory)

According to this theory, the six days of Genesis refer to what God created long after He originally created the universe. The six days are literal twenty-four hour days, but God didn't begin those six days until long after He first created the universe. Let me explain. In Genesis One, each day begins with, "and God said, let...," and each day ends with, "and the evening and the morning were the _____ day." Genesis 1:1 begins and ends with no such statements; therefore, the creation of the heavens and the earth is not part of the first day. It came before the first day. The six days of creation begin later with "And God said, let there be light...." Simply put, Genesis 1:1 describes the original creation. This is followed by a gap of time after which God performs the creative acts mentioned in the six literal days of Genesis 1:3-31, the second creation. It may or may not be assumed that God originally created the heavens and the earth by means of the Big Bang over billions of years. However, when it came to creating/shaping/forming the earth and the life on it as it is today, those creative acts were done within 144 hours. This is an essential part of the Gap Theory. When we examine the science of this theory, we will see that it has no problems with the geological record.

B1b. Twenty-Four Hour Day, Old-Universe, Punctuated-Creation (Multiple-Gaps Theory)

This is somewhat of a combination of the last two theories. The "days" are literal days, but there isn't a single gap of time between the original creation of the universe and the final creation of the earth. Instead, there are long gaps of time between each of the six literal twenty-four hour days. The universe can therefore be billions of years old. God initially performed creative acts during the twenty-four hour period of the first day. Then there was a gap of millions or billions of years followed by another twenty-four hour period of creation called the second day. This was followed by another gap of millions or billions of years, etc. The heavens, the earth, and all life were created in six literal twenty-four hour days, but nothing in the Bible says they had to be consecutive days. This theory has the same two problems with geology as the Day-Age Punctuated Creation Theory has. It doesn't explain living fossils and the order of the geological layers is not correct.

B2. Twenty-Four Hour Day, Young-Universe, Single-Creation (Young-Earth Theory)

This theory teaches that God created the universe, the earth, and life on earth in six literal days. Over a period of 144 hours, God started and finished all the works of creation. This is the Young-Earth Theory. It has historically been the most straightforward interpretation, and probably the most accepted, but it fell into disfavor as scientific evidence seemed to indicate the universe was billions of years old. However, it has staged a comeback, and is quite popular nowadays. It is defended by creationists who believe the universe is only a few thousand years old. Originally, this theory had the same problem with the geological strata. The order of creation did not match the order of fossils found in nature. Initially, that was no problem. The early proponents of this theory believed that fossils and strata were merely geological oddities, created by God, that had no historical or biological significance. That idea has been rejected and replaced by Flood-Geology. Flood-Geology is the theory that all the geological strata and all the fossils were laid down by the Great Flood of Noah's day. Today, Young-Earthers are Flood-Geologists by necessity. If Flood-Geology falls apart, then the Young-Earth Theory falls apart. I believe Flood-Geology falls apart. The Young-Earth/Flood-Geology Theory creates many more problems than it fixes. While I don't believe this theory, I am first to admit that is believed by many, many good Christians.

C. Days of Relativity

This is an interesting explanation of the six days of creation. According to the principles of relativity, the passage of time depends on your frame of reference. Thus, an astronaut traveling at near-light speed could go to a distant star and back in only a few years of his time, but hundreds or even thousands of years would pass on earth. The amount of time it takes the astronaut to complete his trip depends on which frame of reference is used. A thousand years in one frame of reference might be a day in another frame of reference, and this is the key to understanding the days of Genesis. This theory says that the creative days were literal twenty-four hour periods, but only in the Creator's frame of reference, not in the creation's frame of reference. While billions of years passed in the universe's frame of reference, only six twenty-four hour days passed in God's frame of reference. This sounds very scientific, but there are two problems with this theory. The first is that the descriptions in Genesis are given from an earthly perspective, thereby forcing us to accept an earthly frame of reference. The other problem is that God is eternal and has no single frame of reference for time and space. He is not limited or restricted by space and time. He is omnipresent and omnitemporal. As such, He experiences all time-space frames simultaneously. (Actually "experiences" is not the right word to describe God's relationship with time-space, but I don't know how else to describe it.)

D. Days of Revelation

This theory says that God revealed the creation story to Moses over a six day period and then rested from His revelation on the seventh day. According to this theory, there weren't six days of creation; there were six days in which

God revealed what He had created. In other words, God met with Moses over a six day period to reveal the details of creation. It took six days for God to give an account to Moses about what He had created in the beginning. (Creation itself could have taken billions of years.) On the third day, for example, God gave an account to Moses that He had created vegetation, plants, and trees on the earth. God didn't create or make vegetation on the third day; He only revealed this to Moses on the third day of their time together. This idea stems from an interpretation of Genesis 2:4

Gen. 2:4 "This is the account of the heavens and the earth when they were created, in the day that the LORD God made earth and heaven." (NASB)

The word translated as "account" is actually plural. It should say, "These are the accounts..." We're told to believe that the six days of are days of accounts, not days of creation. This would be an acceptable explanation if it didn't contradict Scripture. Exodus 20:11 and other passages clearly tell us that the six days refer to what God created and made.

Exo. 20:11 "For in six days the LORD made the heavens and the earth, the sea and all that is in them, and rested on the seventh day; therefore the LORD blessed the sabbath day and made it holy." (NASB)

Exodus 20:11 doesn't say, "In six days the Lord gave an account of making the heavens and the earth..." It says that the Lord made the heavens and the earth in six days.

E. Days of Divine Decrees

This is another interesting idea. According to this interpretation, God made six great decrees before time began.

Each decree began with "Let there be…" or the equivalent statement. These decrees were made over a six day period so that one decree was made per day. Since God is eternal and omnipotent, as far as He was concerned, making the decree was as good as actually creating the thing He decreed. Since the decrees took six days, God views creation as being six days. The actual formation of the universe could have been any length of time. Regrettably, this view contradicts itself. If these decrees were made before time, then there was no such thing as six days. "Six days" measures a passage of time, but if time wasn't yet created, days couldn't pass, and it couldn't have taken God six days to make the decrees. Further, there is nothing indicating that God is talking about anything else but twenty-four hour earth-days. He's not talking about Jovian days or Plutonian days. He's talking about days from an earthly reference. It would be hard to have earth-days if the earth wasn't yet created.

F. Repeated-Days

Those who believe this theory think there were only three creative "days" and not six. This theory doesn't exactly say what "day" means. It could be short or it could be long. According to this theory, day four is a parallel description of day one; day five parallels day two; and day six is the same as day three. Their rationale is that day one and day four both describe the creation of light. Day one says, "Let there be light," while day four says, "Let there be lights in the firmament…" Day two and day five describe what God did with the waters. Day two describes God separating the waters from the waters, while day five describes God letting the waters bring forth sea creatures. Finally, day three describes the emergence of dry land while day six describes the creation of animal life on that dry land. One of the primary reasons this theory was proposed is because plants were created on day three but the sun wasn't created until day four. Proponents find it hard to believe that plants could exist before the sun, especially if these are day-ages of billions of years. Making day four the same as day one alleviates this problem. This parallelism theory sounds good, but it also contradicts Exodus 20:11. Whatever the "days" were, there were six of them. To its credit, it fits the geological record a little better. It describes the sun, moon, and stars existing before plant life, and aquatic life coming before terrestrial life.

The Other Three Variables

The theories listed above are the major theories dealing with whether the "days" are long periods of time, literal twenty-four hour days, or something else. What follows are some of the theories dealing with the other questions about creation. Remember, these other variables are independent of the meaning of day. Sometimes people will make it seem like you have to accept one particular meaning of day before you can accept some of these other variables. This isn't true. If you will take the time to meditate on these ideas, you will understand how you can blend many of them with almost any definition of day. Just taking into consideration the four questions and the variables I listed, you can come up with 108 creation permutations. We won't do that.

II. THE MEANING OF "CREATE"

A. Co-Eternal Matter and Energy

This idea says that Genesis describes the formation of the universe and the earth, but not their actual creation. Matter and energy, time and space, the fundamental particles and the fundamental forces of nature were already present. In fact, they were eternally present when God began to work on them to fashion a cosmos out of chaos. According to this view, "create" means to form out of preexisting matter and energy. Creation in this sense describes the new things God made out of eternally preexisting matter and energy. Any interpretation of "day" could fit with this theory. We can describe God as working on co-eternal matter and energy over billions of years to form the universe. We can also describe God as quickly working on co-eternal matter and energy to form the universe in six days. This theory runs counter to what cosmologists and astrophysicists tell us about matter and energy. Since the universe is expanding and its entropy (thermodynamic disorder) is increasing, it cannot have been here for eternity. Simply put, nothing natural can be eternal.

B. *Ex Nihilo* Creation

The Latin phrase *ex nihilo* (out of nothing) describes how everything was created out of absolutely nothing by an eternal, supernatural God. Proponents of this theory say there was no preexisting matter and energy. This theory can also be blended with the any definition of "day." God could have gradually created everything out of nothing over a period of billions of years, or He could have quickly created everything out of nothing during six twenty-four hour days.

III. THE MEANING OF "BEGINNING"

A. Long Period of Time

This view says that "the beginning" refers to God's creation of the universe over a long period of time. God initially created the raw materials but put nothing into shape; nothing was complete. At first, there were no stars, planets, etc. Creation started with the creation of time and space, matter and energy. This was followed by the creation/formation of the stars, the galaxies, the sun, the earth, and all the rest. Genesis 1:1 encompasses God's creative acts up to the point of the earth being without form, and void. This theory is almost always combined with the Day-Age Theory because it fits extremely well with the Big Bang Theory. It can also fit the Gap Theory.

B. Instantaneous Point of Time

This view says that "the beginning" was an instantaneous point of time. Genesis 1:1 describes an instantaneous creation of the universe. What the universe contained is not specified. Some say the stars, the sun, etc. Others say nothing but space and the earth. They say the earth was created instantly, but it was created without form, void, and covered by water. Genesis 1:3-31 describes what God did after the initial creation of an otherwise complete universe. This theory is usually combined with the Young-Earth Theory. It absolutely does not fit with the Big Bang Theory, but it could fit with the Gap Theory if the initial creation was instantaneous.

IV. THE MEANING OF GENESIS 1:1 (WHEN COMPARED TO GENESIS 1:3-31)

A. Title

This theory says that, "In the beginning God created the heavens and the earth," is merely the title. Genesis 1:1 is not the first part of creation. Neither is Genesis 1:2. It is a subtitle. Genesis 1:2 describes what earth was like when God began to create. A good example of this is seen in *The Moffatt Bible*. James Moffatt moved the verses around so that the first part of Genesis is a title or introduction.

"This is the story of how the universe was formed. When God began to form the universe, the world was void and vacant, darkness lay over the abyss"

In other words, Genesis 1:1 is not an act of creation. The earth was already without form, and void when God began to form the universe. So, rather than being an act of creation, Genesis 1:1 and 1:2 are a title and subtitle for the acts of creation that follow. This idea doesn't really say what happened, or how God created things. It only states that Genesis 1:1 is a title. This idea can be combined with either the Day-Age Theory or the Young-Earth Theory. It blends nicely with the Co-Eternal Matter Theory, but it can also be combined with The *Ex Nihilo* Theory if God had previously created everything out of nothing before Genesis 1:1 but didn't tell us about it in Scripture. This idea does not fit with the Gap Theory.

B. Summary

This view holds that Genesis One describes creation, but the text of Genesis 1:1-2 is a summary of the events that follow. In other words, Genesis 1:1-2 and Genesis 1:3-31 are parallel descriptions. They describe the same creative acts but with a different emphasis. Genesis 1:1-2 is the summary and Genesis 1:3-31 are the details. There is a biblical problem with this idea. The summary doesn't agree with the details, but I'll talk more about that later. As it stands, this sub-theory can be combined with either the Day-Age Theory or the Young-Earth Theory and with either the *Ex Nihilo* Theory or the Co-Eternal Matter Theory. Genesis 1:1-2 could be a summary of what God did over billions of years, or Genesis 1:1-2 could be a summary of what God did over six days. The earth could have been created without form, and void, or the earth could have eternally been without form, and void. This idea doesn't fit the Gap Theory either.

C. Creative Act

The thought here is that Genesis 1:1 was a creative act or acts that chronologically came before the creative acts of Genesis 1:3-31, before the six days. Genesis 1:1 doesn't say when it happened or how long it took, but it was a separate work of God. This was followed in time by His other acts of creating, making, and forming. The six days describe subsequent creative acts. Both the Young-Earth and the Day-Age Theories try to incorporate Genesis 1:1-2 into the six days of creation. They want to make Genesis 1:1-2 part of the first day, but that doesn't work. As I mentioned, Genesis 1:1 doesn't start with, "And God said let." It is not part of the first day. If God created the earth in Genesis 1:1, and it says He did, then Genesis 1:1 is the initial creation of the earth. What follows in Genesis 1:3-31 does not describe the original creation of the earth. Genesis 1:3-31 does not describe the earth's first beginning. It describes a subsequent beginning; the earth had already begun before Genesis 1:3. Even if it was created in a formless and void condition, it still began before Genesis 1:3. The earth was already created before God said, "Let there be light." This being the case, then Young-Earth and Day-Age creationists must admit there was a gap of time between Genesis 1:1 and Genesis 1:3. Young-Earthers will say it was a gap of minutes or hours while Day-Agers could allow the gap to be billions of years. In spite of this, Day-Agers denounce the Gap Theory. I've never been able to figure out how they can say that. When you ask them when life on earth was created, they will answer that it was about nine billion years after the universe began. This seems to be a fairly long gap of time to me. Young-Earthers are a little more consistent, but they still can't say Genesis 1:3-31 describes earth's original beginning. Genesis 1:3-31 describes what God did to the earth that already began in Genesis 1:1. So how can they object to the idea of earth having two beginnings? It's no longer a matter of if there was a gap of time. It's only a matter of how long that gap was.

My Agenda

As you can see, creation theories can be quite complex. Fortunately I'm not going to investigate all the sub-theories and their various contortions. Instead, I will focus on the four characteristics that define the Gap Theory. If I can show that these are true, then I think the Gap Theory rises to the top of the creation theories. Here are the four characteristics of the Gap Theory.

1. THE MEANING OF "DAY": Twenty-Four Hour Day, Old-Universe, Dual-Creation
2. THE MEANING OF "CREATE": *Ex Nihilo* Creation
3. THE MEANING OF "BEGINNING": A Long Period of Time
4. THE MEANING OF GENESIS 1:1: A Creative Act

I also want to reveal some of the shortcomings of the Day-Age and Young-Earth Theories when they are sifted through the filters of science and Scripture. If these theories fail to pass these tests, then they should be rejected. For instance, if Scripture proves that the six days were real days, then the Day-Age Theory is wrong. If science proves the universe is old, then the Young-Earth Theory is wrong. In addition, I absolutely must show how the Gap Theory resolves the apparent conflicts between science and Scripture better than the other theories. If a theory has a problem with resolving these apparent conflicts, then we can't put much faith in that theory. The Day-Age and the Young-Earth Theories don't solve a lot of conflicts. In fact, they often create bigger conflicts than they resolve.

I now want to examine some problems that seem to arise when we compare the biblical account of creation with the scientific account of creation. **THIS IS THE HEART OF THE MATTER: THE BIBLICAL ACCOUNT OF CREATION SEEMS TO DISAGREE WITH THE SCIENTIFIC ACCOUNT OF CREATION.** We need to see how the different creation theories explain this. When we do that, I think you will see how the Gap Theory is superior to both the Day-Age and the Young-Earth Theories. Both of those theories fail to pass through the filters of true biblical revelation and true scientific observation.

Chapter Three:

Apparent Problems

—⟋⟍⟋—

Scientific Problem 1: You're Out of Order

If the six days in the first chapter of Genesis represent a chronological account of God's creation, and they certainly seem to be exactly that, then a problem arises when we compare the order of creation of organisms with their order of appearance in the geological record. The order in which the Bible says things were created doesn't match the order in which geology says they appeared. The geological strata DO NOT FIT the Genesis account of creation. Let me show you. Look at the order of appearance of things according to the Bible.

Biblical Order of Appearance

Day 6

      Man
      Land Animals
      Land Mammals

Day 5

      Birds
      Marine Animals
      Marine Mammals

Day 4
- Stars
- Moon
- Sun

Day 3
- Fruit Trees
- Seed Plants
- Land Plants
- Dry Land

Day 2
- Atmosphere, Clouds

Day 1
- Light
- Ocean

Now look at the order of appearance of these things according to scientific observations.

Scientific Order of Appearance

(MYA=Million Years Ago; BYA=Billion Years Ago)

0 MYA	Man
50 MYA	
75 MYA	
100 MYA	Marine Mammals
125 MYA	Fruit Trees
150 MYA	Birds
175 MYA	Seed Plants
200 MYA	
225 MYA	Land Mammals
250 MYA	
275 MYA	

300 MYA	
325 MYA	
350 MYA	
375 MYA	
400 MYA	Land Animals
425 MYA	Land Plants
450 MYA	
500 MYA	
550 MYA	
600 MYA	Marine Animals
650 MYA	
700 MYA	
750 MYA	
800 MYA	
850 MYA	
900 MYA	
950 MYA	
1.0 BYA	
2.0 BYA	
3.0 BYA	Ocean
3.5 BYA	Atmosphere, Clouds
3.8 BYA	Moon
4.0 BYA	Dry Land
5.0 BYA	Sun
10 BYA	Stars
15 BYA	Light

Here is a table of information from my college biology textbook, *Biology*.[2] It shows the geological ages and the order of appearance of the various groups and types of plants and animals, both aquatic and terrestrial.

Evolution and the Geologic Ages				
From: *Biology* 2nd Edition by John W. Kimball Copyright 1969 Addison-Wesley Publishing Co.				
ERAS	**PERIODS** / *EPOCHS*	**AQUATIC LIFE**	**TERRESTRIAL LIFE**	**STARTING** (millions of years ago)
Cenozoic	Quartenary / *Recent* / *Pleistocene*		Man in the New World / First Men	0.5
Cenozoic	Tertiary / *Pliocene* / *Miocene* / *Oligocene* / *Eocene* / *Paleocene*	All Modern Groups Present	Hominids and Pongids / Monkeys and Ancestor of Apes / Adaptive Radiation of Birds / Modern Mammals and / Herbaceous Angiosperms	63 +/- 2
Mesozoic	Cretaceous	Modern Bony Fishes / Extinction of Ammonites, / Pleiosaurs, and Icthyosaurs	Extinction of Dinosaurs, / Pterosaurs, Rise of Woody / Angiosperms, Snakes	135 +/- 5
Mesozoic	Jurassic	Pleiosaurs and Icthyosaurs / Abundant / Ammonites Again / Abundant / Skates, Rays, and Bony / Fishes	Dinosaurs Dominant / First Lizards: Archaeopteryx / Insects Abundant / First Angiosperms	180 +/- 5
Mesozoic	Triassic	First Pleiosaurs and / Icthyosaurs / Ammonites Abundant at / First Rise of Bony Fishes	Adaptive Radiation of Reptiles / (Thecodonts, Therapsids, / Turtles, Crocodiles, First / Dinosaurs) / First Mammals	230 +/- 10
Paleozoic	Permian	Extinction of Trilobites / and Placoderms	Reptiles Abundant / (Cotylosaurs, Pelycosaurs) / Cycads and Conifers; Ginkgoes	280 +/-10
Paleozoic	Pennsylvanian	Ammonites, Bony Fishes	First Reptiles / Coal Swamps	310 +/- 10
Paleozoic	Mississippian	Adaptive Radiation of / Sharks	Forests of Lycopsids, / Sphenopsids, and Seed Ferns / Amphibians Abundant / Land Snails	345 +/- 10
Paleozoic	Devonian	Placoderms, Cartilaginous / and Bony Fishes / Ammonites, Nautiloids	Ferns, Lycopsids, and / Sphenopsids / First Gymnosperms and / Bryophytes / First Insects / First Amphibians	405 +/- 10
Paleozoic	Silurian	Adaptive Radiation of / Ostracoderms; / Eurypterids	First Land Plants (Psilopsids, / Lycopsids) / Arachnids (Scorpions)	425 +/- 10
Paleozoic	Ordovician	First Vertebrates / (Ostracoderms) / Nautiloids, Plinia, Other / Mollusks / Trilobites Abundant	None	500 +/- 10
Paleozoic	Cambrian	Trilobites Dominant / First Eurypterids, / Crustaceans, Mollusks, / Echidnoderms	None	600 +/- 50

		Sponges, Cnidarians, Annelids Tunicates		
Pre-Cambrian		Fossils Rare But many Protistan and Invertebrate Phyla Probably Present	None	

Now, let me remove the dates and simply compare the orders of appearance.

Biblical Order vs. Scientific Order

Man	Man
Land Animals	Marine Mammals
Land Mammals	Fruit Trees
Birds	Birds
Marine Animals	Seed Plants
Marine Mammals	Land Mammals
Stars	Land Animals
Moon	Land Plants
Sun	Marine Animals
Fruit Trees	Ocean
Seed Plants	Atmosphere, Clouds
Land Plants	Moon
Dry Land	Dry Land
Atmosphere, Clouds	Sun
Light	Stars
Ocean	Light

Don't look too closely and things may seem to fit. Many Christian Creationists gleefully state that the scientific order of appearance matches with the biblical order of appearance. They say this because there are some points of agreement. Both accounts show that man is the pinnacle of God's creation. Both accounts say that dry land came before land plants and land animals, and both say that the oceans came before marine life. (Amazing!)

The problem becomes apparent, however, if you look at the details. If the biblical order matches the scientific order, then we would be able to draw lines from one side of the chart to the other and have nothing but parallel, uncrossing lines. If we draw lines from one side to the other, and any lines cross, then the biblical account doesn't match the scientific account. Here is what we get when we connect the lines according to the order of appearance.

Biblical Order vs. Scientific Order

Biblical Order	Scientific Order
Man	Man
Land Animals	Marine Mammals
Land Mammals	Fruit Trees
Birds	Birds
Marine Animals	Seed Plants
Marine Mammals	Land Mammals
Stars	Land Animals
Moon	Land Plants
Sun	Marine Animals
Fruit Trees	Ocean
Seed Plants	Atmosphere, Clouds
Land Plants	Moon
Dry Land	Dry Land
Atmosphere, Clouds	Sun
Light	Stars
Ocean	Light

As you can see, the two accounts don't match. When two lines cross, it means there is a discrepancy between the biblical account and the scientific account, and there are lots of crossed lines, aren't there? Look at the appearance of the moon for example. The biblical order of the appearance of the moon disagrees with the scientific order in respect to the appearance of the stars, fruit trees, seed plants, land plants, the oceans, and the atmosphere and clouds. That's six points of disagreement. The biblical order of appearance of fruit trees doesn't match the scientific order of appearance of the sun, the stars, marine animals, land animals, land mammals, and birds. That's six more points of disagreement. Another major disagreement is that the Bible appears to place the water-covered earth even before the presence of light. That would mean the water-covered earth existed before the Big Bang. There are other disagreements.

THE BIBLE TEACHES THAT THE EARTH WAS CREATED COMPLETELY SUBMERGED IN WATER AND THAT DRY LAND APPEARED LATER

According to the Genesis 1:2 and 1:9, the entire earth was under water before dry land was present. If you study the history of the earth according to science, you find no such condition. In fact, you find the opposite situation. According to cosmologists and geologists, the earth was formed either by the accumulation of space debris or by an explosion of the sun. In either case, the violent forces would have caused earth to be a molten mass. Only as it cooled, could a crust (dry land) form. Even after the crust formed, it still would have been hundreds of degrees for millions of years. Oceans couldn't have formed until the crust cooled to a point that allowed liquid water to collect. According to the scientists, dry land appeared long before there was surface water. The Bible and science seem to disagree on this point.

GEOLOGY REVEALS THAT BIRDS CAME AFTER LAND ANIMALS WHILE THE BIBLE TELLS US BIRDS CAME FIRST

According to the Bible, birds were created on day five. Land animals weren't created until day six. The geological record reveals that land animals came before birds. Something seems wrong. Is the biblical record wrong? Is the geological record wrong? They both can't be right about this, can they? If birds came after land animals, then the Bible appears to be wrong. If birds came before land animals, then the geological record appears to be wrong.

GEOLOGY REVEALS THAT WHALES CAME AFTER LAND ANIMALS, BUT THE BIBLE SAYS THEY CAME BEFORE

Again we see an apparent conflict between what the Bible says and what science reveals. Geology shows that land animals came first. According to the Bible, the great creatures that live in the water, which includes whales, were created first. Whales were created on day five but land animals weren't created until day six. If whales came first, the Bible is right but geology is wrong. If land animals came first, geology is right but the Bible is wrong. Something doesn't seem to match.

GRASS AND TREES WERE CREATED BEFORE MARINE LIFE

The Bible says God created grass, trees, herbs, and seed-bearing plants on day three. The Bible also says God created sea creatures on day five. According to the Bible, trees and seed-bearing plants came before life in the oceans. Geology firmly teaches the opposite. According to the geological evidence, the oceans teemed with life long before any land plants; especially before seed-bearing plants, flowering plants, and fruit-bearing trees.

LAND PLANTS COME BEFORE THE SUN

Here we have another contradiction in theories. The Bible says that God made land plants on the third day but that He didn't make the sun until day four. You can't blend this with science. According to cosmologists and astronomers, the sun was present for billions of years before land plants. If the sun was created before land plants, then day four came before day three. That, of course, would mean that God made a mistake.

I think you get the idea. There are even more apparent discrepancies than the ones I've listed. Look back at all the lines that cross in the diagram and you'll see other problems. The Bible and science seemingly disagree on the appearance of land animals vs. seed plants, marine animals vs. land plants, birds vs. marine mammals, and others. **No wonder unbelievers laugh at the Bible. The Bible seems to contradict science right out of the starting block.** Whatever creation theory we come up with, it needs to explain these apparent contradictions, and it needs to explain them without creating bigger biblical and scientific problems. Both the Young-Earth and the Day-Age Theories actually create more biblical and scientific problems than they solve. One problem with the Day-Age Theory is that it tries to reconcile the Genesis account with the fossil record because it assumes the fossils are a physical record of creation in Gen. 1:11-31.

Scientific Problem 2: Missing "Missing Links"

The fossil record is very important to both Day-Age Creationists and to evolutionists, but for different reasons. Day-Age creationists assume the fossils show the geologic ages in which God created the various life forms. For them, the fossils provided a record of what God had **created** during the six Day-Ages. For evolutionists, however, the fossils provided a record of what had **evolved** over the long ages. About the only thing in common between the Day-Age Theory and the Theory of Evolution was the belief that species had very gradually changed over long periods of time. Day-Age creationists insisted that God had done this. Evolutionists insisted that Darwinian evolution had done this. Both theories agreed that life changed very gradually over very long periods of time. Unfortunately for both theories, this wasn't what the fossil record revealed. The fossil record didn't seem to provide the "missing links" needed by both theories. Let's first look at how the fossil record affected the Theory of Evolution.

According to Darwin, species changed in very small, very gradual steps, over very long periods of time. Although Darwin had no "missing links" that connected any of the biological groups together, he believed that someday they would be discovered. Almost everyone in the scientific community agreed. A crucial point to remember is that Darwin and the evolutionists didn't know how biological traits were produced or what caused biological traits to change. They also didn't know how traits were passed down to following generations. No one knew about DNA at the time. They simply believed that for some unexplained reason, biological traits changed very gradually over time. They also believed that local environments changed very gradually over time. So, it made sense to them that once a very gradual biological change occurred, it would very gradually influence an organism's survival in its very gradually changing environmental niche. Gradualism was the paradigm of the day. This is where the term "survival of the fittest" came into play. Darwinian evolutionists asserted that there were two forces driving evolution forward. The first force produced change in heritable biological traits. The second force was the constantly changing environment. An organism with a trait that caused it to survive would be more likely to pass that trait on to the next generation. An organism with a trait that caused it to die wouldn't do as well. Over very long periods of time, species would very gradually change according to their ability to survive in the ever changing environmental niches. Those organisms with traits that helped promote survivability would very gradually replace those organisms that didn't have those traits. If what they asserted was true, then the fossil record would show very gradually changing biological traits as time passed. In fact, the changes in traits would be so gradual that eventually even the Linnaean Classification System would break down. The traits of one species would gradually blur into the traits of another species. One genus would have traits that imperceptibly changed into the traits of another genus. One family's traits would change in such small gradations that it would be impossible to draw a definitive line where it changed into another family. The same would be true for order, class, phylum, and even kingdom. Alas, the fossil record didn't reveal what they wanted.

I like explaining things with illustrations because sometimes it makes difficult concepts easier to grasp. Imagine that you were an artist and you wanted to paint a mural of the visible color spectrum. You would start on the left side of your canvass with red and work your way to the right with orange, then yellow, then green, then blue, then purple, and finally back to red. Let's also imagine that your canvass was a 100 miles long. It's a big project. You would start with a can of paint that had 1,000,000 parts of pure red paint and nothing else. With that paint, you would paint a one-inch wide line from the top of the canvass to the bottom of the canvass. Next, you would get a can of paint with 999,999 parts of red paint and 1 part of orange paint. With that paint, you would draw another one-inch line. This line would be touching the first line and be parallel to it for its entire length. Then you would get a can of paint with 999,998 parts of red paint and 2 parts of orange. You would use this to paint your third one-inch line. For your fourth line you would use paint that was 999,997 parts red and 3 parts orange. This process would continue until you painted one million one-inch lines. At that point you would have a can of paint with 1,000,000 parts of orange paint and no red paint. Once you drew the pure orange line, you would go to your next can of paint. It would contain 999,999 parts of orange paint and 1 part of yellow paint. Very gradually, line by line, you would have one more part of yellow paint and one less part of orange paint in your paint cans. Once you completed the next million lines, you would have a can of pure yellow paint. Then you would use the paint that was 1 part green and 999,999 parts yellow. You would continue this process color by color. You would add more and more green to your yellow, then more and more blue to your green, then more and more purple to your blue, and finally more and more red to your purple. Each can of paint would differ from its predecessor by one part per million. The total transition through all six colors (red, orange, yellow, green, blue, and purple) would require six million one-inch lines. Your color spectrum would evolve through six million steps in the space of about 100 hundred miles.

Once you rested from your work, what would you expect to see? You would expect to see a rainbow that changed so gradually and in such small increments that you wouldn't be able to tell where red ended and orange began. You wouldn't be able to distinguish the 100% blue line from the 99.9999% blue/0.0001% purple line. You would be able to classify the colors into two "kingdoms," the hot colors (red, orange, and yellow) and the cool colors, (blue, green, and purple) but you wouldn't be able to tell when one "kingdom" ended and the other "kingdom" began. You would be able to see all six "phyla," (red, orange, yellow, green, blue, and purple) but you would find it difficult to see the lines of demarcation. You could create an entire classification scheme by labeling the colors as "blue-green," "bluish green," "greenish blue," or "greenish blue-green," but it would be hard to decide when something was blue enough to be called "bluish blue-green" instead of "greenish blue-green." The end result would be that your classification scheme would be inadequate; you wouldn't be able to fit things into neat little compartments. This is the kind of results you would expect from Darwinian evolution except it would be even harder to define species.

In our rainbow mural example we have six million steps in 100 miles. In the case of living organisms, we are dealing with billions of DNA changes over hundreds of millions of years. Changes would indeed be very, very, very gradual! Darwinians insisted that the fossil record would eventually reveal this kind of evolution. They believed that enough fossils would be discovered so that it would be plain for all to see that Darwinian evolution explained the origins of species. They should have found billions of transitional fossils.

Is this what they found? No; a century of intense searching passed and they kept finding the same kind of fossils, of the same kind of organisms, with the same kind of traits, in the same kind of strata. In fact, this is how the Fossil Index System was made possible. The Fossil Index System was a way of determining when an organism lived based on its position in the fossil record. Because they kept finding the same kind of organisms in the same geological strata all over the world, they assumed that the fossil record was a chronological record of worldwide evolution. When they discovered particular fossils with particular traits, they were able to place them into a particular chronology. Each stratum had its own characteristic fossils with their own characteristic traits. But if Darwinian evolution was true, then one would not expect to keep finding the same kind of organisms with the same kind of traits in the same kind of strata all over the world. Different species with different traits in different environments would display such gradual transitions that you wouldn't be able to tell when a fossil index started and when it ended. **The Fossil Index System seemed to indicate that there had been abrupt changes in biologic traits as well as abrupt changes in environments**. Evolutionists realized that if Darwinian evolution was true, then the changes from one species to the next would be so gradual that there should be nothing but transitional fossils in the strata. It began to appear as if biological traits weren't gradually changing. They kept finding Devonian strata with Devonian fossils, Permian strata with Permian fossils, Jurassic strata with Jurassic fossils, etc. The very fact that these groupings kept appearing seemed to indicate that evolution could be grouped into categories. This was a great discovery for the Fossil Index System, but it was a bad discovery for Darwinian Evolution. Evolutionists couldn't find the "missing links."

Of course, many evolutionists claimed they found the "missing links," but those discoveries always proved to be insufficient evidence because no one could unquestionably prove they were gradual transitions from one major group to another. Oh, they might find an organism that had a trait that appeared similar to the traits in other groups but the transitions were still too large and still too abrupt to be universally accepted as proof, even by other evolutionists. You see, it wasn't just creationists who began refuting Darwinism. By the mid-twentieth century, a number of staunch evolutionists also began doubting Darwin's model. Why did they doubt? For starters, it was discovered that biologic traits were the products of genetic information stored in strands of DNA. The only way to change biologic traits was to change the information. The only way to get new information in DNA was by random mutations. Because random mutations constantly occur and because environmental niches are constantly changing, there should be no true fixity of species. Nothing should stay the same for very long. **Finding species that remained constant should be the exception and not the rule.** Finding the same kind of fossils, with the same kind of traits, in the same geological strata, all over the world was an awkward situation for gradualism. There were large gaps between the different organisms and they couldn't find many (if any) truly transitional fossils. If Darwin was right, the geological strata ought to show an uncountable number of transitional fossils with imperceptibly small changes. Unfortunately, they couldn't find that. This eventually became such an embarrassment for evolutionists that they quietly laid Darwin to rest and created another Theory of Evolution. That theory is called Punctuated Equilibria. Simply stated, evolutionists now claim that evolution didn't happen very gradually over very long periods of time. Instead, it happened in very short periods of very rapid changes, with very long "quiet" periods of time in between. Very rapid bursts of evolution are separated by very long interludes of no evolution. Because the changes were so rapid and the time periods were so short, there would be very few transitional organisms. That means there would be even fewer transitional fossils. Fish evolved to amphibians so quickly that it was unlikely any transitional forms would be found in the fossil record. Reptiles evolved to birds just as quickly. In fact, one evolutionist stated that a reptile laid an egg and a bird hatched out. Not much chance of finding transitional fossils in that case. This new theory could now be "proven" because it was based on the lack of evidence. The more transitional fossils you didn't find, the more "true" this theory became. Of course, real science is never strengthened by relying on the idea that the less evidence there is, the more scientific it becomes. Yet, this is what evolutionists have done over the past few decades.

So, how are gaps in the fossil record a problem for creationists? Well, they're not a problem for some creationists. Young-Earth creationists have no problem explaining this. God simply did not create life forms out of previous life forms. Gap Theory creationists agree. But, the Day-Age Theory initially said that God very gradually created newer species from older species in very gradual steps over very long ages of time. If God had created life this way, then the fossil record should reveal billions of gradual transitions over hundreds of millions of years. Initially, Day-Age creationists believed that since God gradually created over long periods of time, then the fossil record would reveal very small, very gradual changes over time. In essence, they believed the fossil record would be identical to what the evolutionists were hoping for. Those creationists who believed this theory couldn't explain the gaps in the fossil record any better than the Darwinian Evolutionists. If God created life this way, then there should be uncountable numbers of "created" transitional forms among the fossils. Unfortunately for Day-Agers, their theory had a scientific problem: the missing links were missing. There were too many gaps in the fossil record. What could they do? Well, it wasn't long after the new Evolutionary Punctuated Equilibria Theory was formulated that the new Creationist Punctuated Day-Age Theory was formulated. I'm sure this was a coincidence. Punctuated Day-Age creationists now say that God didn't create new creatures gradually over long periods of time. Instead, He periodically created new species suddenly. Like the Punctuated-Equilibria crowd, the Punctuated Day-Age crowd uses the lack of transitional fossils as their "proof." Yes, it explains the gaps in the fossil record, but it still fails to explain the discrepancies in the order of appearance. It also fails to explain "living fossils."

Scientific Problem 3: Living Fossils
(Yea, Though They Once Were Dead, Yet Shall They Live)

According to evolutionists and Day-Age creationists, the fossil layers are a record of the history of the appearance of life on earth. This was why gaps in the fossil record were so embarrassing for evolutionists. The fossil record kept ruining their theory. Nevertheless, evolutionists put a lot of stock in The Fossil Index System. The system assumed that simple organisms in the lower strata evolved into more complex organisms in the upper strata. Likewise, Day-Age creationists put a lot of stock in The Fossil Index System.

They assumed that God created simple organism first and more complex organisms later. The Fossil Index System wasn't a bad system because each specific stratum seemed to contain specific organisms regardless of where on earth the stratum was found. Nevertheless, there were problems with the system. One problem was that scientists kept discovering "living fossils." These were organisms that according to the Fossil Index System had died out millions of years ago, but were found to be alive today. The Coelacanth fish was a perfect example. This "primitive" fish was supposed to have died out seventy million years ago. They were found in strata older than seventy million years, but then they disappeared from later strata. So it was assumed that Coelacanth became extinct seventy million years ago. Day-Age creationists were okay with this. God had created these "primitive" fish, and then they died out seventy million years ago. Then in 1939 fishermen caught a live Coelacanth off the coast of Acapulco. Many have been caught in other parts of the world since. **The fossil record did not provide an accurate history of the Coelacanth fish.** The same was true for numerous other living fossils that have been discovered. Examples are the Tuatara lizard, the Metasequoia tree, the Ginkgo tree, Neopilinia (a small marine mollusk), Lepidocaris (a crustacean), and over a hundred others. According to the fossil record all these things were shown to have been extinct for millions of years. Yet, all these things are living today. Evolutionists either had to admit that evolution was wrong or that the fossil record was less reliable than thought. They admitted the latter of course, but in doing so they cast doubt on the entire Fossil Index System.

Day-Age creationists face the same problem. **The fossil record is not a reliable record of when God created things during the six long Day-Ages.** According to Day-Age thinking, God created Ginkgo trees 200 million years ago during day three of creation. The fossil record showed they died out even before land mammals appeared. This means they died out before day five. So, if they died out before day five but are alive today, then God must have gotten His days mixed up. He must have created trees after day five, or else Ginkgoes wouldn't be here. But if He did that, then the Bible is wrong when it says God created trees on day three. If you try to fit all the different "living fossils" into which day they were created, you get a terrible mess. Neopilinia was created on day five with the other marine animals. Their fossils, however, show they became extinct before seed plants appeared on day three. This means they became extinct even before they were created. Then somehow they reappeared alive and well after being dead for 400 million years. The Day-Age Theory doesn't explain "living fossils" without having God get confused about what He created on which day. The Gap Theory has a much better explanation.

Chapter Four:

Day-Age Solutions

—ɯ—

The most glaring problem Christian creation-scientists face is that the biblical order of appearances doesn't seem to match the scientific order of appearances. The Day-Age Theory doesn't solve the problem. **Making the "days" long periods of time doesn't change their order.** Day-Agers can say the geological strata represent the six long ages, but they still must admit that the ages are in the wrong sequence. According to them, God doesn't know the meaning of "first," "second," "third," etc. Apparently God doesn't know the meaning of "day" either.

The Meaning of "DAY"

What does the Bible reveal about how we should translate the word "day" in Genesis One? I believe we can discover its meaning if we simply examine how the word is used here and in other places in the Scriptures. I believe we can eliminate all creation theories that interpret "day" to be symbolic, to be an age, or to be anything other than a regular twenty-four hour earth day. Remember, God is describing things from an earthly perspective.

YOWM is the Hebrew word for "day." What does it mean? When used in reference to time, *YOWM* can define three different periods.

1) It is used for the daylight portion of a twenty-four hour day.
2) It is used of a complete twenty-four hour day.
3) It is also used to describe an undetermined length of time.

Day-Agers contend that the six days of creation cannot be literal twenty-four hour days. They insist that "day" is equivalent to "age." They say that Genesis One is God's way of symbolically describing what He did during the billions of years between the Big Bang and the creation of man. Day-Agers are quick to defend their belief by telling us that *YOWM* can mean a long period of time. This is possible, although that's not its usual meaning. Nevertheless, because of this possibility, they feel they are justified in believing the days are long ages. Unfortunately for them, this possibility becomes impossible when you let the Bible interpret itself.

So what does *YOWM* mean in Genesis One? The answer is found in Genesis One. You see, God defined the meaning of "day" in the text. God tells us exactly what period of time *YOWM* describes. *YOWM* is first used in Genesis 1:5. Note that *YOWM* is used twice in this verse with a different meaning each time.

Gen. 1:5 "And God called the light **DAY**, and the darkness He called night. And there was evening and there was morning, one **DAY**." (NASB)

In its first usage in Genesis 1:5, *YOWM* was used as the name for a period of light. "God called the light (*OWR*) day (*YOWM*)." This means *YOWM* and *OWR* are equivalent; they described the same period of time. "Day" in this context is a period of light. I don't know how God could have made this any more clear. *YOWM*, whatever its length, was a period of light. It was not a period of light-and-dark, light-and-dark,

light-and-dark, as is required by the Day-Age Theory. This is
obvious since God also used "night" (*LAYIL*) to describe the
period of time when it was dark (*CHOSEK*). This addition
of *LAYIL* is important because it narrows down the definition of *YOWM*. Used in this way, *YOWM* means the daylight
portion of a twenty-four hour day. Day-Agers fail to look at
the word "night." Yes, *YOWM* can be used for a long period
of time, but *LAYIL* is never used for a long period of time.
LAYIL is used 233 times in the Bible and it is always the dark
portion of a twenty-four hour day when referring to time. In
the Genesis account of creation, *LAYIL* is paired with *YOWM*
indicating that *LAYIL* and *YOWM* represent equal periods of
time. It makes no sense to speculate that "day" means billions of years when it is combined with the word "night"
that always means the dark hours of a twenty-four hour
day. Furthermore, the words "day" and "night" (*YOWM* and
LAYIL) are combined in 54 Biblical passages. In every other
portion of the Bible, when these words are used together
to define a period of time, they refer to a twenty-four hour
period. Why would the Hebrew be translated differently in
Genesis One? This causes a difficulty for those who defend
the Day-Age Theory. The Bible doesn't describe what God
did during six days. The Bible describes what God did during
six days AND six nights. The addition of the word "night" is
a big problem for those who believe the six days are six ages.
YOWM does not mean "age" in the first part of Genesis 1:5.

    YOWM is used a second time in Genesis 1:5. God says,
"And there was evening and there was morning, one *YOWM*."
Day-Agers ignore the words "evening" and "morning"
(*EREB* and *BOQER*). God says that *YOWM* had an evening
and *YOWM* had a morning. He said the same for the second
through sixth days. The Genesis account of creation plainly
shows that *EREB* and *BOQER* describe the same period of
time as *YOWM*. What do *EREB* and *BOQER* mean?

EREB is used 132 times in the Bible and it means the short period of time we call evening, dusk, or twilight. *EREB* does not refer to a long period of time. When used in reference to a measurable period of time, the Bible never uses *EREB* to describe anything but the twilight hours of a twenty-four hour day. What grammatical basis is there for saying *EREB* means a short period of time everywhere in the Bible except in the first chapter of Genesis?

The same is true for *BOQER*. It is used 205 times in the Bible and it refers to sunrise, dawn, or the beginning of the light portion of a twenty-four hour day. Again, when used to describe a measurable period of time, it always means a short period of time. It is never used for a long period of time in the rest of the Bible. What gives anyone the right to give it a unique meaning in Genesis One?

EREB and *BOQER* are combined in 44 Biblical passages. Every time these words are used together, they refer to an ordinary day. Why would "evening and morning" have a different meaning in Genesis One?

The Day-Age defenders look at the word "day," but not its modifiers. The first chapter of Genesis uses a cardinal number for the first day ("one") and ordinal numbers ("second," "third," "fourth," etc.) for the other days. This is significant. It's as if God anticipated the "Non-Literal Day" crowd and wanted to make it clear to them that He is talking about ordinary days. After saying *YOWM* had a period of light and a period of dark; after saying it had an evening and a morning, He then gives the MEASURABLE LENGTH OF TIME it encompassed. It was "one day." God gives no hint that He is talking about long ages. If He meant long ages, then He didn't express His thoughts very well. He used words that mean "one literal day" everywhere else in the Bible. When an ordinal or cardinal are combined with *YOWM*, the meaning becomes specific. Throughout the entire Bible, when these seven numerical modifiers are added to the Hebrew word for "day" it means a twenty-four hour period.

In my studies of the biblical account of creation, I have discovered that it doesn't take much effort to find conflicting opinions among the scholars. There are Hebrew scholars who will agree with what I just said. They agree that when one of these numerical modifiers is added to *YOWM*, it always refers to a literal day. On the other hand, there are scholars who are just as intelligent and just as educated and just as qualified, who say that this isn't so. They say that a numerical modifier combined with *YOWM* can mean an unspecified time period in Hebrew. The hard part for me is that I really admire and respect some of the scholars in the second group. I would like to tell you how wrong they are and how ignorant they are, but I have read some of their works and know that they are wonderfully brilliant Christians. So how do I explain the fact that I think they are absolutely wonderful but absolutely wrong? I can only assume that they base their opinion on extra-biblical Hebrew writings. Apparently *YOWM* plus a number doesn't have to mean a twenty-four hour day when you look at the entire history of the Hebrew language. While this may be true in other writings, I still insist that in the Bible, *YOWM* plus a number always refers to a literal day; at least that's the case when we are talking about the six days of creation and the seventh day of rest. Words and word usages change over time and geography so I can't discount that what they say is true, but it doesn't apply to God's Word. I know this sounds pretty arrogant coming from someone who is not a scholar, but you don't have to be a scholar to determine the truth about *YOWM*. All you need is an exhaustive concordance of the Hebrew Old Testament and a lot of time on your hands. If you don't have access to an exhaustive Hebrew concordance, or the time, don't fear. I've done the work for you. Here are the biblical usages of *YOWM* when modified by the cardinal number one and the ordinal numbers two through seven.

(*ECHAD YOWM*)

First Day [5]	Gen. 1:5	Ezra 10:16	Ezra 10:17
Neh. 8:2	Hag. 1:1		
One Day [18]	Gen. 27:45	Gen. 33:13	Lev. 22:28
Num. 11:19	1 Sam. 2:34	1 Sam. 27:1	1 Ki. 20:29
2 Chr. 28:6	Ezra 10:13	Esther 3:13	Esther 8:12
Isa. 9:14	Isa. 10:17	Isa. 47:9	Isa. 66:8
Zec. 3:9	Zec. 14:7		

You can find other instances of "first day" in the KJV, such as in Lev. 23:24. Actually it says, "on the first of the month," not, "on the first *day* of the month." The word "day" is in italics and not in the original Hebrew. When *ECHAD* and *YOWM* are plural, the translation is rendered, "a few days." This is seen in Gen. 27:44, Gen. 29:20, and Dan. 11:20. There are some other things to consider. For instance, Ex 12:15 and Ex 40:2 both contain "first day" in the KJV, but *ECHAD* is not used in these verses. *YOWM* is used, but they use a different word for "first." In addition, some manuscripts have *ECHAD* in Ezra 3:6 while other manuscripts apparently don't. I found it listed in some translations but not in others. There are a few other minor variations, such as "same day," "each day," "a day," and "daily," when *ECHAD* and/or *YOWM* are used with other numerical modifiers. However, they always refer to a real twenty-four hour day; never to an extended or unspecified period of time. The first day of creation was one day; not an age!

(*SHENIY YOWM*)

Second Day [13] Gen. 1:8   Exo. 2:13   Num. 7:18
Num. 29:17   Josh. 10:32   Judg. 20:24   Judg. 20:25
1 Sam. 20:34   Neh. 8:13   Esther 7:2   Jer. 41:4
Eze. 43:22

As is the case of "first day," there are verses that are translated "second day of the month," in the KJV when the word "day" is in italics and not in the original. Two such examples are 1 Sam. 20:27 and 2 Chr. 3:2. Various other modifiers are used with *YOWM* that result in a translation of "next day," but *SHENIY* is not used. Again, in all of these usages, *YOWM* still refers to a regular day. The second day of creation was one day; not an age!

(*SHELIYSHIY YOWM*)

Third Day [31] Gen. 1:13   Gen. 22:4   Gen. 31:22
Gen. 34:25   Gen. 40:20   Gen. 42:18   Exo. 19:11
Exo. 19:16   Lev. 7:17   Lev. 7:18   Lev. 19:6
Lev. 19:7   Num. 7:24   Num. 19:12   Num. 19:12
Num. 19:19   Num. 29:20   Num. 31:19   Josh. 9:17
Judg. 20:30   1 Sam. 30:1   2 Sam. 1:2   1 Ki. 3:18
1 Ki. 12:12   1 Ki. 12:12   2 Ki. 20:5   2 Ki. 20: 8
2 Chr. 10:12   2 Chr. 10:12   Esther 5:1   Hosea 6:2

Like the case of "one day" and "second day," there are some cases of "third day" that don't use the word *YOWM* and there are some other cases where other words are used for "third" instead of *SHELIYSHIY*. Even with those differences, they always refer to a regular twenty-four hour day. The third day of creation was one day; not an age! Wait! Day-Agers point to Hosea 6:2 as an exception to the rule.

They say that Hosea 6:2 proves that *YOWM* can mean an age. Let's look at Hosea 6:2.

Hosea 6:1-3 "Come, let us return to the LORD. He has torn us to pieces but he will heal us; he has injured us but he will bind up our wounds. {2}After two days he will revive us; on the **third day** he will restore us, that we may live in his presence. {3} Let us acknowledge the LORD; let us press on to acknowledge him. As surely as the sun rises, he will appear; he will come to us like the winter rains, like the spring rains that water the earth." (NIV)

Day-Agers point out that the restoration of Israel did not take place within three literal days. Therefore, an ordinal before *YOWM* can mean an extended period of time. I disagree! Look carefully what the text reveals. Hosea demands that Israel first return to the Lord. Hosea says that if Israel would truly acknowledge the Lord, then He would restore it within three days. Israel had only to acknowledge God and by the third day it would be restored. The problem wasn't with God. The problem was with Israel. The people didn't denounce their evil. They didn't acknowledge the Lord. They didn't return to the Lord. The restoration did not come within three days because Israel failed to do as Hosea demanded. Hosea gives no hint that God is talking about anything but literal twenty-four-hour days. No symbolism is hinted. His wording in the text includes such concrete images as the sun rising, the winter rains, the earth, and the spring rains. As you read the rest of the chapter you see that Hosea mentions the early dew, flashes of lightning, and the morning fog. Nothing in this chapter is symbolic of anything. Rather than being symbolic, God is announcing that He will respond to repentance immediately. "As surely as the sun rises," He will respond. He will not postpone His response if Israel returns to Him. It's blasphemy to think that God would ask His beloved people to acknowledge Him, and then after they do, make them wait three billion years to be restored. Rather than proving these days are long periods of time, Hosea proves that God is willing to work in very short periods of time to bless His people. God told Hosea that He was ready to restore Israel within seventy-two hours. True, the restoration of Israel did not come within three literal days, but the sad thing is, it could have. (As a point of history, however, this passage was fulfilled. After three days in the tomb, Christ restored those Jews who acknowledged Him.)

What is true of days one through three is also true for days four through seven. All these numerical modifiers of *YOWM* refer to regular twenty-four hour days.

(REBIY'IY YOWM)

Fourth Day [6]   Gen. 1:19   Num. 7:30   Num. 29:23
Judg. 19:5   2 Chr. 20:26   Ezra 8:33

(CHAMIYSHIY YOWM)

Fifth Day [4]   Gen. 1:23   Num. 7:36   Num. 29:26
Judg. 19:8

(SHISHSHIY YOWM)

Sixth Day [6]   Gen. 1:31   Exo. 16:5   Exo. 16:22
Exo. 16:29   Num. 7:42   Num. 29:29

(*SHEBIY'IY YOWM*)

Seventh Day [47 or 48]		Gen. 2:2	Gen. 2:2
Gen. 2:3	Exo. 12:15	Exo. 12:16	Exo. 13:6
Exo. 16:26	Exo. 16:27	Exo. 16:29	Exo. 16:30
Exo. 20:10	Exo. 20:11	Exo. 23:12	Exo. 24:16
Exo. 31:15	Exo. 31:17	Exo. 34:21	Exo. 35:2
Lev. 13:5	Lev. 13:6	Lev. 13:27	Lev. 13:32
Lev. 13:34	Le. 13:51	Lev. 14:9	Lev. 14:39
Lev. 23:3	Lev. 23:8	Num. 6:9	Num. 7:48
Num. 19:12	Num. 19:19	Num. 19:19	Num. 28:25
Num. 29:32	Num. 31:19	Num. 31:24	Deu. 5:14
Deu. 16:8	Josh. 6:4	Josh. 6:15	Judg. 14:17
Judg. 14:18	2 Sam. 12:18	1 Ki. 20:29	Esther 10:1

There is one more *SHEBIY'IY YOWM* I didn't list; Judges 14:15. I didn't list it in the above group because it is controversial. Ah-Ha! Finally here is a real controversy! If you are a Day-Ager and you want to use a Bible verse to destroy all I have presented, then here is your verse. This is the one verse that might be used to show that "seventh day" does not mean "seventh day." How can you use this verse to prove that "seventh day" doesn't mean "seventh day?" Well, for some reason some of the translations say "fourth day" instead of "seventh day."

Judg. 14:15 "And it came to pass on the **seventh day**, that they said unto Samson's wife, Entice thy husband, that he may declare unto us the riddle, lest we burn thee and thy father's house with fire: have ye called us to take that we have? *is it* not *so?*" (KJV)

Judg. 14:15 "Then it came about on the **fourth day** that they said to Samson's wife, 'Entice your husband, that he may tell us the riddle, lest we burn you and your father's house with fire. Have you invited us to impoverish us? IS this not *so*? (NASB)

I've looked at several translations and there seems to be a fairly even split over what Judges 14:15 tells us. Is it "seventh" or is it "fourth?" I don't know. Of course, even if you decide to blast my arguments by claiming "seventh day" can mean "fourth day," you're still stuck with the fact that it means an ordinary day. "Day" still doesn't mean "age" in this context.

ECHAD YOWM is used 23 times in the Old Testament. *SHENIY YOWM* is used 13 times. *SHELIYSHIY YOWM* is used 31 times. *REBIY'IY YOWM* is used 6 times. *CHAMIYSHIY YOWM* is used 4 times. *SHISHSHIY YOWM* is used 6 times. *SHEBIY'IY YOWM* is used either 47 or 48 times. God gives no clues anywhere in the Bible that these are something other than real twenty-four hour days. In every other portion of the Bible, when these modifiers are attached to "day," it refers to a twenty-four hour period.

The Bible clearly describes the first *YOWM* as a regular day. It does the same for the second *YOWM* and the third *YOWM*. Now, even if I tried to close my eyes and pretend I didn't see God's clear definition of *YOWM* for the first three days, I could no longer pretend that *YOWM* was an age by the time I got to the fourth *YOWM*. Why? It's because the Bible mentions a greater light and a lesser light on the fourth *YOWM*. What were these two lights? They were the sun and the moon. On *YOWM* four, God made the sun the dominant source of light during the *YOWM* and He made the moon the dominant source of light during the *LAYIL*. Even if *YOWM* was a long period of time for the first three days, it couldn't be a long period of time for the last three days. By adding this information about the sun and the moon, we cannot mistake God's intent. God is talking about regular days and regular nights. This creates a conflict with the Day-Age Theory. Up to this point, the only life God had created (restored) was land vegetation. It was on days five and six that God created life in the oceans and animal life on the land. Day-Agers contend that "days" five and six had to be a period of about 400 million years. God says it was two days. I'll leave it to you to decide whom you want to believe.

Yes, *YOWM* by itself can mean an age of time. However, when *YOWM* is used in context with modifiers and companion words that have a specific meaning, then *YOWM* takes on that same specific meaning. This is exactly what we see in our language. Let me give you an example. Let's say I looked out my window and told you, "There's a pig in my garden." Without looking out my window, what image comes to mind? Well, you'd probably think of a four-footed animal with a short snout and a curly tail. That's the most common meaning of the word "pig." Ah, but in the 1960's, "pig" was a derogatory word for "policeman." Couldn't someone argue that I was looking at a police officer in my garden? Yes, even though "policeman" is not the usual definition of "pig," such an argument could be valid. Now, if I said, "There's a pig and a cow in my garden," wouldn't it make the definition of "pig" be a little more specific? Of course it would. As soon as I added a companion word with such a specific meaning, it would be difficult to think that I was telling you there was a police officer and a cow in my garden. Furthermore, if I said, "There's a pig and a cow in my garden; and I'm talking porcine and bovine," then it would be impossible to mistake my meaning. Finally, if I said, "There's a small animal called a pig and a large animal called a cow in my garden," then only an idiot would think I was talking about a police officer.

This is exactly what God does with the word *YOWM*. Its most common meaning is an ordinary day. To make this meaning clear to us, He adds the specific companion word, *LAYIL*. He then adds two more companion words, *BOQER* and *EREB* to make His meaning more exact. Finally, to make His meaning impossible to mistranslate, He adds two more companion words, "light" (*OWR*) and "darkness" (*CHOSEK*). The creation account in Genesis cannot be subjected to twisted interpretations. "Night" always means "night." "Morning" always means "morning." "Evening" always means "evening." All of these words refer to portions of the normal twenty-four day. By adding the words "light" and "dark," God makes it evident that He's talking about real days and nights. By referring to them as one day, and as the second day, and the third day, on up to the seventh day, He makes it evident that He's talking about a real seven-day week. By mentioning the sun in the day and the moon at night means these are not ages. It is glaringly obvious that "day" can only be translated in the same way its companion words are, and that is a literal day. Clearly, six literal days are not compatible with the Day-Age Theory. So who's wrong; God or the Day-Agers?

If the Six Days are Ages, What are the Six Nights?

If the Day-Age Theory is correct, then the six day-ages of creation span thirteen billion years. This interpretation of "day" creates another problem. According to Day-Agers each "day" on the average was about two billion years long. But according to the Bible each day had ONE day and ONE night. Each day had ONE morning and ONE evening. Each day is said to consist of a single period of dark and a single period of light. The Bible says that each day had an evening (*EREB* in the singular) and a morning (*BOQER* in the singular). It doesn't say that each day had evenings and mornings. Each day had a single period of light and a single period of dark. Each single light period followed each single dark period. God used the singular forms to describe each period of light and each period of dark. The Hebrew can't be translated to indicate that each day had billions of mornings and billions of evenings unless the plural forms of *EREB* and *BOQER* were used. God didn't use the plural forms. Since there were six instances of the singular forms, then there were six singular periods of light and six singular periods of dark. If each "day-age" consisted of a single period of light (day) and a single period of dark (night), then we would be forced to believe the absurd notion that each light period would have been a billion years long and each dark period would have been a billion years long. Now, what would happen to living things if the earth was subjected to periods of total darkness lasting a billion years? What would happen to living things if the earth was subjected to periods of total light lasting a billion years? The earth would be a frozen ball of ice for a billion years and then it would be molten for a billion years. No life form could withstand a billion years of constant darkness or a billion years of constant light. Now I admit that no one, not even Day-Agers, believe that the periods of light and periods of dark lasted a billion years. Yet once you accept the notion that each "day" had billions of days and billions of nights, you have to wonder why God said that each day had only one day and one night. The Day-Age Theory doesn't fit well through the filter of Scripture.

If the "DAYS" are Long Ages, Then How Did Certain Plants Reproduce?

As I have mentioned, the Day-Age Theory has a problem with the order of strata seen in the geological layers. Making the "days" into long ages doesn't correct the incorrect order of appearance. Besides not correcting the geology problem, it creates an even bigger biology problem. If the Day-Age Theory is true, then the "days" of Genesis One were extremely long periods of time. If the "days" represent billions of years, then a problem arises with the survival of many species. There are numerous species of plants that require the presence of animals in order to reproduce. Sometimes it's species specific. There is a species of Yucca plant, for instance, that requires a particular species of Yucca moth to carry its pollen. Only the Yucca moth has the ability to do this. Certain fig trees need certain fig-wasps to do their pollinating. Thousands of flowering species require some kind of animal to pollinate them. The seeds of many seed-bearing plants must pass through the digestive tracts of animals to be softened before they can germinate. All of these plants, created on day three, would have had to wait billions of years before the necessary animals were created on days five and six. How could they have survived billions of years without being able to reproduce?

The Really Big Problem: Day-Age Weakens the Gospel

In addition to creating more scientific problems, the Day-Age Theory creates a bigger biblical problem. According to Hebrew scholars it is very risky to interpret a word or phrase in a unique fashion if it is defined throughout the Bible in a different way. There is a danger in accepting a unique definition as a way of defending your favorite doctrine. Once you do it, you have no basis for objecting to someone else's unique definition as a way of defending their favorite doctrine. To defend the Day-Age Theory from a grammatical basis, you have to accept these unique meanings of words:

1) *LAYIL* would have to have a unique meaning.
2) *EREB* would have to have a unique meaning.
3) *BOQER* would have to have a unique meaning.
4) The combination of *YOWM* and *LAYIL* would have to have a unique meaning.
5) The combination of *EREB* and *BOQER* would have to have a unique meaning.

6) The combination of a numerical modifier and *YOWM* would have to have a unique meaning.

If we allow these words to mean something other than what they plainly mean, then we find ourselves with some real problems in trying to understand the Bible's account of creation. Now as bad as it is to have a theological problem with creation, an even bigger theological problem arises if we permit this same kind of liberty with word usage elsewhere in the Bible.

Jesus said He would be in the grave three days and three nights (Matthew 12:40). What if someone said these were three ages? On what grammatical basis could you prove they were wrong? If you insisted that "day" in Genesis meant an age, how could you say they were wrong if they insisted that "day" in Matthew also meant an age? You couldn't! You'd have no right to say they were wrong. Their New Testament Day-Age Theory would be just as valid as your Old Testament Day-Age Theory, but what would that mean to Christianity? It would mean that Jesus is not alive today. It would mean He is still dead and in the grave. It would mean His disciples and followers didn't really see Him alive after the crucifixion. It would mean that everything we believe as Christians is a lie. It would mean the Gospel accounts are not true. It would mean that Paul's epistles are based on a lie. None of the New Testament could be considered reliable or historical. It would mean the New Testament writers were lying to us. Dear, dear Christian friends, it is a dangerous thing to allow man's opinions to overshadow God's teachings. We are in real danger of turning the resurrection of Christ into a meaningless yarn if we can't trust God to define the meaning of day. The six days of creation in Genesis must be regular twenty-four hour days. If "day" can mean something else in the creation accounts, then "day" can mean something else in the resurrection accounts. If that is possible, then we have no defense for the visible, physical resurrection of Our Lord three days after His crucifixion.

Chapter Five:

Young-Earth Solutions

—m—

The geological record and the biblical record don't match, or at least they don't seem to match. Now, I hope you're astute enough to see where I'm going with these geological problems. The difficulty is that most people think the geological strata and the six days of Genesis are records of the same event. They assume that geology is the PHYSICAL RECORD of the six days of creation and the Bible is the WRITTEN RECORD of the six days of creation. But, what if they aren't the same event? What if the geological strata are a historical record of something other than the six days of Genesis? If this were so, then we wouldn't necessarily expect the biblical order of appearances to match the geological order of appearances. There are two ways this problem can be resolved.

The first solution is Flood-Geology. If Young-Earth Flood-Geologists are correct, then the geological strata have nothing to do with the order of creation in Genesis One. Regardless of the sequence of creation, things got jumbled up during the Flood. The six days of Genesis record the order of creation, but geology records the results of The Flood. This is why Young-Earthers must also be Flood-Geologists. Otherwise, they have to fall back on their original belief that the fossils and geological strata are merely strange and exotic rock formations that have no connection with living things. If the geological strata were not caused by The Flood, then Young-Earthers have the very same problems with geology that the Day-Agers have. Young-Earthers must be Flood-Geologists, or their theory gets washed away.

The second solution is the Restoration Theory. I believe the Gap Theory resolves these problems in a better way. The geological strata record the original creation (Genesis 1:1) while the six days of Genesis record the restoration (Genesis 1:3-31). The orders of appearance don't match because they are two different creations. Geology is a physical record of the first beginning and the six days of Genesis are a written record of the second beginning. Earth has had two beginnings. More on that later; let's now look at the Young-Earth Theory.

The Young-Earth Theory

Many say that the Young-Earth Theory was never challenged until the coming of the age of science. This isn't so. Long before our modern scientific age, people wondered about the age of the earth. For as long as historical records have shown, there have been different cosmologies, religions, and philosophies trying to answer why and how and when we got here. Christians aren't the only people who have wondered about the age of the earth. The problem is that humanity hasn't been able to agree on what source of information is reliable. Do we trust the ancient pagan religions? Do we trust the Chinese calendar? Do we trust the ancient Greek philosophers? Do we trust science? How about the Mayan calendars? Where do we find truth? As Christians, we know we have THE source of truth, but that source of truth doesn't actually tell us when God created the heavens and the earth. This may shock you, but it is true. The Bible does not say! There is no statement in the Bible that says God created the universe on any particular date. All creation chronologies derived from the Bible are biblical INTERPRETATIONS, not biblical DECLARATIONS. The most famous of these interpretations is the 1650 Ussher Chronology. James Ussher

(1581-1656) was the Archbishop of Armagh (Northern Ireland). By looking at the genealogies in the Old Testament and comparing them with some known historical events, Archbishop Ussher decided that the universe was created on October 23, 4004 B.C. There have been others who have done similar chronologies and have arrived at other specific dates, but Ussher's date has long been the accepted view by many Christians. After all, it seems pretty reasonable that you could look at the genealogies ("so-in-so" begat "so-in-so") in the Old Testament and calculate the age of the earth. It's not that easy. The reliability of this dating technique has been disputed for as long as it has been proposed. First, it is argued that "begat" in the Old Testament doesn't have to mean being the direct father of someone. It can mean grandfather, great-grandfather, great-great-grandfather, etc. Second, it is argued that there are gaps in the names mentioned in the same gene-alogies when comparing one portion of Scripture to another. In other words, these aren't necessarily direct father to son lists. Jesus is called the son of David, but the time between David and Jesus was about a thousand years. It's tempting to exploit this idea, but I'm not going to get into this argument. At best, it is a weak argument. Even if there are gaps, I don't think we could get millions of years from Adam to me. It still seems we would be looking in the range of thousands of years. I will, however, point out one problem with Ussher's conclusion. Deriving dates from the genealogies can only determine the time of the creation of Adam; not the creation of the heavens and the earth. Ussher first assumed there was no gap of time between Genesis 1:1 and Genesis 1:3. The Gap Theory and the idea that the universe was very old were discussed long before Ussher was born. In fact, as we will see later, the Gap Theory predates the Christian Church. Ussher rejected an old earth out of hand. His logic was based on cir-cular reasoning. He assumed the earth was young, (no long gap of time before Adam) and then since his conclusion was that the earth was young, he believed his original assumption was correct. This kind of chronology doesn't disprove the Gap Theory because it is based on the assumption, not on the evidence, that the Gap Theory is untrue. It doesn't prove the earth is young; it merely ignores the possibility of the earth being old.

Why is this important? Because Young-Earth creationists still use this same faulty logic. They insist that the Gap Theory can't be true because the genealogies "prove" the earth is young. Their "proof" doesn't prove anything because it is circular in its reasoning. They don't seem to understand that the gap came before the genealogies and has nothing to do with the genealogies. They use another faulty "proof" against the Gap Theory as well. They say that since the days of Genesis are six literal days, the earth must be young. Again, they bring circular reasoning to their defense. Six literal twenty-four hour days don't prove the earth is young unless you first assume there was no long gap of time before the six literal twenty-four hour days. Let me give an event from my life to show you how they err.

One summer I decided to put a new roof on my house. Being somewhat handy with tools, and desiring to save myself a bunch of money, I decided I would do it myself. I tore off the old shingles, laid down new tar paper, and hammered all the new shingles into place. I must say I did a very nice job, too. In fact, I put more care into the project than most professional roofers would have; it was my house! It was a job done right and a job done well. Now, here's my question:

It took me seven days to complete this project. In what year did I re-roof my house?

You can't know can you? The fact that it took seven days, and these were literal twenty-four hour days, doesn't tell you when it happened. There is insufficient data to be able to determine a date. The only way that seven days could be used to calculate a date would be if you knew seven days from a pre-given date. Without a pre-given date, you wouldn't know if it was last year or ten years ago. I would have to supply you with a starting date before you could calculate when it happened. If I didn't give you a starting date, then a seven day time period doesn't solve anything. The Bible doesn't give us a starting date for the six days of creation. The fact that it was six literal days doesn't tell us when it happened. It doesn't prove the earth is young unless you first know there was no gap of time before the six days started. Young-Earth creationists assume there was no gap of time before the six days, and based on that assumption, they conclude that their assumption is correct. This is circular reasoning on a par with the circular reasoning of evolutionists. (By the way, I did my roofing project during the summer of 1981... I thought some might want the answer.)

The Bomb Squad

There are other flawed arguments the Young-Earthers use against the Gap Theory. Over the years, I have seen Young-Earth creationists repeatedly try to destroy the Gap Theory by using a particular "biblical bombshell" that "proves" they are right. Let me defuse this "bomb." Young-Earth creationists insist that the Restoration Theory is wrong because it means death reigned before Adam. Gap Theory creationists look at the geological strata and say that all those fossils (dead things) were formed before Adam was created. Young-Earth creationists become indignant when they hear this. They believe there was no death before Adam sinned. "By one man sin entered into the world, and death by sin," they'll quote. According to them, Romans 5:12 and 1 Corinthians 15:21 prove there was no death before Adam sinned. If you question them, however, they usually admit that some kind of death must have existed before Adam's fall. Plants died when eaten. Cows didn't carry microscopes to pick off the microscopic mites living on the blades of grass before they ate. These mites would have been killed by the cow's digestion system. Insects, nematodes, and other soil-living creatures got squished as great herds of buffalo crossed the plains. Plankton-feeding whales would have digested microscopic animals along with microscopic plants. Once something is digested, it's pretty good and dead. The outer-most layer of epithelial tissue (skin) consists of cells that are dead. If those skin cells were alive, we would be in excruciating pain every time something touched us, and worse than that, we would dehydrate within hours. The layer of dead epithelial cells acts as a fluid barrier to prevent desiccation. If those cells weren't dead, we soon would be! Furthermore, those dead cells are constantly being rubbed off and replaced by the growing cells beneath them. Epithelial cells aren't the only cells continually dying in order for animals to live. Red blood cells have a constant turnover rate. White blood cells do the same. Osteoclasts are cells that continually kill bone cells so that new bone cells can be made by osteoblasts. If this didn't happen, bones couldn't grow. Baby animals couldn't become adult animals if living cells didn't die. If you study embryological development, you'll discover that certain embryonic cells have to die in order for further development to take place. Without death, animals couldn't have multiplied according to God's command. There had to be death of some kind, even in animals.

Many Young-Earth creationists take this no-death theory one step farther and say the Second Law of Thermodynamics didn't exist before sin. (The Second Law of Thermodynamics teaches that things tend to become more disorderly over time as energy is expended.) Here is one example of how their solution creates a ridiculous scientific problem. They equate the increase in universal entropy (disorder) with sin. They believe that since the earth was "very good," there was no Second Law of Thermodynamics. But, if that were true, then Adam couldn't have walked and his heart couldn't have pumped blood. Cells need energy provided by oxidative biochemical reactions in accordance with the Second Law of Thermodynamics. Without increasing entropy, plants couldn't move water up their roots. Without increasing entropy, plants couldn't make sugar by the process of photosynthesis. In fact, without the Second Law of Thermodynamics, the sun couldn't shine. Yes, God could create a miraculous economy not dependent on the laws of physics, but that was not the condition of the earth in Genesis 1:2. The earth was already in a high entropic condition of great physical disorder. The Second Law of Thermodynamics wasn't created because of Adam's sin. **The rejection of a scientific law is a strange way to defend a religious belief.** Besides, if there had been no death, how could Adam have understood when God warned him not to eat the fruit of the Tree of Knowledge of Good and Evil? If there had been no death, Adam wouldn't have known what God was saying. It would be like me telling you, "If you don't finish reading this book, I am going to skitzlebirger you." This is a nonsense word and conveys nothing. Why would God warn Adam with nonsense he couldn't understand? It makes God look rather foolish. However, the real answer to their criticism is not to speculate about possible errors in their theory, but to see if the Bible really says what they claim.

Rom. 5:12 "Wherefore, as by one man sin entered into the world, and death by sin; and so death passed upon all **MEN**, for that all have sinned:" (KJV)

1 Cor. 15:21-23 "For since by man *came* death, by man *came* also the resurrection of the dead. *{22}* For as in Adam all die, even so in Christ shall all be made alive. *{23}* But every

MAN in his own order: Christ the firstfruits; afterward they that are Christ's at his coming." (KJV)

The key point they overlook (sometimes in ignorance and sometimes on purpose) is that these passages are talking about the death of MEN, not about death in general. "Passed upon all MEN" and "every MAN in his own order," mean we are not talking about the death of animals and plants. Applying this to all living things means that all the Brussels sprouts my mother ever cooked will someday receive resurrection bodies and live in heaven with Christ. After all, it says, "in Christ shall **ALL** be made alive." Is this ridiculous? Yes, of course. The Bible does not say there was no death before sin. The Bible says there was no death for MAN before sin. God did not intend for man to die. They were the ones made in His image, not animals. This is a big distinction. In addition, the Bible doesn't say that God was going to kill animals if Adam sinned. God gave no warning to Adam that his sin would lead to the death of all living things.

Gen. 2:17 "but you must not eat from the tree of the knowledge of good and evil, for when you eat of it **YOU** will surely die." (NIV)

God told Adam that he, not the animals, would die if he ate of the tree of the knowledge of good and evil. God gave no warning of universal judgment. Still, Young-Earth creationists claim that Adam's sin affected the entire universe. They quote Romans 8:22.

Rom. 8:22 "For we know that the whole creation groans and suffers the pains of childbirth together until now." (NASB)

This is true, but Romans 8:22 is not Romans 5:12. Romans 5:12 mentions Adam's sin. Romans 8:22 DOES

NOT mention Adam's sin. It tells us that all creation is suffering, but it doesn't attribute it to Adam. Can we blame the suffering of the entire universe on Adam's sin? Did Adam's sin cause Jupiter and Saturn to groan? Is Andromeda Galaxy suffering the pains of childbirth because of Adam? To what extent did Adam's sin reach? The Bible tells us.

Gen. 3:17 "Then to Adam He said, 'Because you have listened to the voice of your wife, and have eaten from the tree about which I commanded you, saying, 'You shall not eat from it'; Cursed is the **GROUND** because of you; In toil you shall eat of it All the days of your life.'" (NASB)

"Cursed is the GROUND because of you." That's all God says. The ground was cursed because of Adam. It doesn't say God cursed the stars and the galaxies and the quasars because of Adam. Young-Earth creationists are adding to God's Word when they claim that all death is the result of Adam's sin.

Them Bones, Them Bones, Them Dry Bones

I would now like to lob a few "bombs" back at the Young-Earthers. I'm not doing this out of meanness. Remember, I was once a Young-Earther myself, so I'm not trying to imply that I'm somehow superior. I do this out of a desire to let the truth glorify our Lord Jesus Christ. I fear that some of what the Young-Earth creationists teach may actually cast doubt on God's character and on the Gospel of Jesus Christ. I can't let this slip by unchallenged. I can excuse a lot of ignorance and misconceptions, since we all have lots of ignorance and misconceptions. However, there are some errors that are without excuse.

A large number of Young-Earth creationists are opposed to the Restoration Theory because it means that God built this present world on top of "a heap of bones," and they are appalled at the idea. They despise the Restoration Theory. They hate the idea of God making something living out of something dead; something good out of something bad. They say that God would never do such a horrible thing. They ask, "How could God say the earth was very good on day six if it was a restoration of something filled with death and decay?" Some claim that God would never take something corrupt and make something good out of it. They go so far as to imply that such an act would violate His Holy character.

Friend, if you have this attitude, then you have just slapped Jesus Christ in the face. If you believe this and claim to be a Christian, then you'd better get down on your knees and reevaluate your relationship with the Lord. This is exactly what Jesus did on the Cross. This is exactly what Jesus did for me, so don't expect me to ignore your insults against His character. I was dead and corrupt and filled with decay, but He went to the Cross to restore me and make something very good of me. I was dead, but now I'm alive. There was nothing good or clean or wholesome in me. Nothing! To claim that God would never restore something corrupt would mean that these Young-Earth creationists despise the work of Jesus. He came to make something good out of something corrupt. Besides, how can you object to God making a perfect world out of a corrupt world when this is precisely what God will do in the future? (2 Peter 3:13)

Taking a kinder and gentler approach, let me say that most Young-Earth creationists don't realize how grievous the consequences of their beliefs are. Some do, yet they cling to them. Regrettably, there is more at stake than whose view of science is better, and that is why I bring up these points. Recall how I objected to the Day-Age Theory because it means we could interpret Christ's time in the grave as three ages, not three literal days. The Young-Earth Theory presents a false idea that allows this same kind of destruction of the Gospel. I'm concerned that some well-known Young-Earth creationists knowingly use false ideas to defend their view of creation. How does this look to unbelievers? **How can an unbeliever believe we represent the God who said, "Thou shalt not lie," when we use lies?** How can an unbeliever believe Christ will take away his corruption and restore him to new life when he's told that God would never restore anything corrupt? When Christian Creationists of any camp teach things that can be used against the character of God or against the message of the Gospel, I become quite concerned.

Apparent Age

This point is so important that it must be fully addressed. This deals with the concept of Apparent Age. What is Apparent Age? Well, Apparent Age is the idea that God created certain things in His universe so that they appeared old, even though they weren't. Now, there is a certain amount of truth to this. When God created Adam and Eve, they appeared to be adults. One day after they were created, they didn't appear to be one-day old babies. The same was true for mighty oak trees. Mighty oak trees were mighty oak trees, not saplings. They had tree rings, giving them the appearance of age. If we could have taken photographs of all the living things in Eden on the day they were created, we would have pictures of plants and animals in all stages of maturity. There would be an appearance of many different ages even though everything was one day old. Baby ducks would be just as old as mama ducks, but they would have had different apparent ages. The question that needs to be addressed is this: Did God create things with apparent age simply for the sake of appearances? Putting it another way, did God create things to appear old just so we would make incorrect conclusions about His creation? Believe it or not, this is the line of thinking that has been used by the "church" for centuries. When early geologists began looking at the strata, they began to realize that the earth appeared old. The "church" came along and said God merely put the strata there to make the earth appear old, even though it wasn't. When early paleontologists realized that fossils made the rocks appear old, the "church" came along and said God merely created fossils to make the rocks look old. There was no connection between fossils and living organisms. Fossils were just strange rock formations created by God. There had never been any plants or animals like those on the earth. When early astronomers began peering out into space, they realized the universe appeared old. The "church" came along and said that God only made it appear that way, and that telescopes were instruments of the devil. I won't go into more detail; you see what I saying. The "church" has earned a bad reputation down through the centuries by its stubborn rejection of scientific observations that disagree with "accepted theology." We do little better today, and Apparent Age is one of those accepted theologies that I fear will drive more people away from the Gospel than draw them to it.

I don't think God created things with apparent age merely to trick us or deceive us. I think He created things the way they were because that was the way they best functioned. Oak trees had rings because tree rings give strength to trees. Mama ducks were bigger because baby ducks needed somebody to lead them to the water. Adam and Eve were adults because as adults they could fully enjoy fellowship with God. God gave everything Functional Age not Apparent Age. Now, trying to ascribe function to things can be difficult. This is especially true when we realize that we're talking about how God views functions. Nevertheless, I cannot find anything in the Bible that indicates God would create things with apparent age merely for the sake of appearances. **The Bible says that we can learn things about God by viewing**

His handiwork, but if His handiwork is skewed, if what we see is not what really is, then what does viewing His handiwork tell us about God? I understand perfectly why an unbeliever wonders about God when he is told that God miraculously created fossils in the strata just to trick him into thinking the earth was more than 6,000 years old. What function can fossils possibly serve? None, other than to prove that God is deceptive.

We need to be very careful when we stand so near such a dangerous precipice. Let's not push logic and reason over the edge and make God a god of deception and fakery. Let's not defend our interpretations of the Bible when such interpretations run counter to well-proven scientific facts. The effects are devastating. God appears to be untrustworthy if our observations of His universe cannot be trusted. The deadly trap is this: **ONCE WE CONVINCE PEOPLE THEY CAN'T TRUST GOD'S REVELATION OF THE PHYSICAL WORLD, HOW CAN WE ASK THEM TO TRUST HIS REVELATION OF THE SPIRITUAL WORLD?**

As we look out into space, we see stars and galaxies that are thousands, millions, and even billions of light-years away. As a scientist, I would look at a star a million light-years away and say that the light I now see actually left that star a million years ago and is just now reaching the earth. I reach this conclusion by using the same logic I use when standing at the train station in New York, watching a train come in from Boston. I know how far away Boston is. I know how fast the train goes. I can calculate when the train left. The Young-Earth Creationist criticizes my logic. I can do that for a train from Boston, but I can't apply that same logic to the light from a distant star. The reason my logic is faulty, he tells me, is that God created the light from that star in transit. In other words, God not only created the star, but He also created light waves emanating from that star. God created the star and He made a beam of light all the way to earth on the fourth day of creation so that it would appear to be in the night sky. Thus, God gave the star Apparent Age.

In 1987 astronomers had the privilege of witnessing a supernova, one of the most awesome displays seen in the heavens. I'm not an astronomer, so I can't describe this in the best astronomical terms, but simply put, a supernova is a star that has burned so much of its mass that it can no longer sustain its size. When this happens, it explodes. The event seen in 1987 was the supernova of a star located in the Large Magellanic Cloud approximately 150,000 light-years from earth. How do we explain what happened? The answer from astronomers is that the star actually exploded 150,000 years ago, and the light of that explosion had just then reached the earth. The star they saw before the explosion had actually been gone for 150,000 years, but it took that long for the last of its light to reach earth. This is the same explanation a Day-Ager or a Restored-Earther would give. A Young-Earther couldn't accept this. If he did, then he'd have to admit the universe is at least 150,000 years old. If the Young-Earthers are correct, then the farthest away we could see a supernova would be 6,000 light-years. The only answer the Young-Earther can give is to say that God had again created an appearance of age. He made it appear as if that star had exploded 150,000 years ago. At this point, however, they step over the edge of logic and take the Gospel with them. Here's why. The star is not there now. The star couldn't have been there 150,000 years ago according to the Young-Earth Theory. The universe didn't exist 150,000 years ago. If God created a star in the Large Magellanic Cloud 6,000 years ago and made it go supernova that very day, it would take 144,000 more years before we could see it explode. So, what astronomers observed in 1987 couldn't have been the explosion of a star in the Large Magellanic Cloud. It was only an apparent explosion of a star in the Large Magellanic Cloud. But how can a star appear to explode and no longer be there if it never really exploded? Since the star is not there now, it means that God never created a star there. He merely created a beam of light so that it appeared as if the star was there. But, if He never created the star, then what astronomers observed in 1987 wasn't the explosion of the star. It was only an apparent explosion of an apparent star. The star never existed. The explosion never happened.

In order to make the universe appear old, God created the appearance of a star and the appearance of an explosion, but neither was real. **If Young-Earthers are right about the age of the universe, then the 1987 Large Magellanic Cloud supernova proves we cannot trust our eyes.** This means we can't tell if something we see is a real something, or if it's merely an apparent something. If I can't trust a truly repeatable, truly testable, truly proven scientific observation, then how can I trust anything else my senses tell me? How am I to know if the freight train speeding my way is a real freight train or simply an apparent freight train? How can I know if I should jump off the tracks? The answer is, I can't! Now, I know that interpretations of observations can be argued, but Christians have no right to argue that the observations aren't real. If I see a rock on the ground, you can't tell me the rock doesn't exist. This is, in effect, what they're saying about supernovae. They say these supernovae never really happened. How ironic is it that some atheists are desperately trying to prove what really happened is what really happened, while some Christians are desperately trying to prove what really happened is what really didn't happen? If Young-Earth creationists are correct, then God has given us an untrustworthy revelation of His universe. If the heavens declare anything about God, they declare that He is deceptive. The works of His hand are a slight of hand. In short, we can't trust God because one of His invisible attributes would be the attribute of deception.

Now, here's where the Gospel comes in. How do we know that Jesus rose from the tomb? We know because there were eyewitnesses. In fact, more than 500 people saw the Risen Christ. Regrettably, some people of other religions argue against the Resurrection by saying that God merely sent them an illusion of a Risen Christ. Jesus was still dead and rotting in the grave, but God sent these 500 people an apparent Christ. How do we respond to such a claim? We reject it on the basis that God is not a God of deception or trickery. We insist that God would never deceive us. BUT HOW CAN WE SAY THAT, IF WE ALSO INSIST THAT GOD MADE APPARENT EXPLOSIONS THAT NEVER HAPPENED OF APPARENT STARS THAT NEVER EXISTED? If the Young-Earthers are right about the age of the universe, then we can't know the Resurrection of Christ was real. Eyewitnesses don't mean squat! It may have been an APPARENT RESURRECTION OF AN APPARENT CHRIST. The bottom line is this: If God can deceive us about a star exploding in the heavens, then God can deceive us about Jesus rising from the tomb. If Jesus didn't rise from the grave, then our faith is in vain.

"Hold on," they'll scream. They can prove how light from a star 150,000 light-years away could have gotten here in 6,000 years. Young-Earth creationists have a Plan B. They say the speed of light is slowing down. They say the speed of light was millions of time faster in the past. Therefore, the light of that exploding star would have gotten here much quicker. Well, I'm going to let the professional physicists and astronomers answer that one. Go read their books. They have mathematical formulas far more complex than my brain can handle. I'm not that smart, so I'll make my argument simple. According to the Laws of Physics, matter and energy cannot be created or destroyed. If the Young-Earth Theory depends on something that violates a proven scientific law, then it ought to be tossed out. It does, and it should!

You see, a well known mathematical formula of physics is Albert Einstein's equation $e=mc^2$. (e is energy, m is mass, and c is the speed of light.) This means that matter can be changed to energy, and energy can be changed to matter, and the quantity of matter or energy involved is determined by the speed of light. If I have a rock with a certain mass, I can calculate how much energy would be generated if the rock was completely transformed to energy. I would do this by multiplying its mass (m) by the speed of light squared (c^2). Now, if the speed of light was greater in the past, then my rock would have had more energy in the past. If the speed of light was a million times greater in the past, then my rock would have had a trillion (a million times a million) times more energy in the past then than it does now. In fact, the total energy of the universe in the past would have been a trillion times greater. This would mean that the universe has lost energy over time. In other words, if the speed of light has slowed down, then energy has been destroyed since the creation. This violates the Laws of Physics. Matter and energy cannot be created or destroyed. This Young-Earth explanation contradicts a proven Law of Physics. It also contradicts the claim that Christ sustains all things (Heb. 1:3). If Christ really is sustaining all things, then the mass and energy of the universe has been sustained. Nothing could be lost. Of course, if you're good with math, you could rewrite this formula as $e/c^2=m$. Then you could claim that the speed of light is slowing down, but energy is not being lost. Instead of energy being destroyed, it would mean that mass is being created. If this is true, then we have a trillion times more mass in the universe now than at creation. Sadly, that violates the same Law of Physics. It also invalidates the claim that God finished creating on the seventh day (Gen. 2:3). Combine Genesis 2:3 with Hebrews 1:3 and you'll see that the Bible told us that matter and energy cannot be created or destroyed. This is another one of those little scientific clues

He gave us. This is how He set up this universe. Matter and energy cannot be created or destroyed. These are the Works of His hands and these are the Words of His lips. They don't contradict. Why are Christians so obstinately defending creation theories that contradict His Work and His Word?

Okay, okay, we'll go to Plan C. Many Young-Earth creationists now talk about how God warps space so that the light from that exploding star could have gotten here faster. It traveled through Riemannian Space. Yeah, try telling that to the traffic cop who pulls you over for speeding. "Honest, officer. I was only doing 35. It's just that my car was going through Riemannian Space so that it appeared I was going 85." What is Riemannian Space? Riemannian Geometry deals with the geometry of curved surfaces. We are probably more familiar with Euclidian Geometry, which is Plane Geometry, Solid Geometry and a few other things dealing with the dimensions of space. You probably remember from high school geometry that one of Euclid's axioms was that the shortest distance between two points is a straight line. It's true, but what if those two points are on a curved plane? Could there be a shorter distance? Yes, most certainly. If you lived in London, England and wanted to go on a jolly holiday to Sydney, Australia, then the shortest distance would be about 10,000 miles. But that's the distance along the curved surface of the earth. If you could bore a hole through the earth, you could shave a couple of thousand miles off your trip. Now, even though it would be shorter by going through the center of the earth, I wouldn't recommend it. It would take a frightfully long time and once you broke through the crust and hit all that magma and molten iron, your holiday wouldn't be very jolly. Well, jolly or not, this is what some very well known and highly respected Young-Earth Creationist use to explain the travel of light from distant stars. The light from all those stars millions and billions of light-years away got here (and is still getting here) via Riemannian Space. It means that photons of light are escaping the "surface" of our universe, traveling into a "place" that is not our universe (not part of the time-space continuum), and then reemerging just so we can see them as stars.

There are so many problems with this idea that I don't know where to begin. First, no one has ever observed Riemannian Space. Science demands repeatable, testable observations. Young-Earthers can't provide it. They use science fiction in place of science fact. Second, for something to escape the confines of the time-space continuum, assumedly it would have to reach a velocity greater than the speed of light. The photons traveling at the speed of light would have to be traveling faster than themselves. Then, once they got out of the universe, they would have to get back into the universe at just the right spot so that they would appear to be coming from just the right place. Amazingly amazing!

There is an even bigger "bomb" that destroys this hypothesis: YOU CAN TEST IT YOURSELF. All you need are two very sensitive, highly complex, superbly engineered sensing devices… your eyes. Go outside on a clear night and look up at the stars. Light from those stars is traveling along the "surface" of our curved universe, Euclidian Space. Supposedly, light is also traveling "straight through" Riemannian Space. Now, if this is true, then every star that is less than 6,000 light-years away should be present in the night sky two times. Alpha-Centauri, the nearest star to our sun is about four and a half light-years away. If its light is traveling through both Euclidian Space and Riemannian Space, then we should see two Alpha-Centauri's separated by a space of however far it has moved in four and a half years. We should be able to see a Euclidian Alpha-Centauri and Riemannian Alpha-Centauri in the night sky. Furthermore, new Euclidian Stars should be popping into view every night. Today, we can see Euclidian Stars that are less than 6,000 light-years away. A thousand years ago, we could have seen only Euclidian Stars that were less than 5,000 light-years away. At the time of Christ, we could have seen only Euclidian Stars that were less than 4,000 light-years away. At the time of King David, astronomers could have seen only Euclidian Stars that were less than 3,000 light-years away. As time passed, the light from farther and farther Euclidian Stars would have finally reached us. When they did, they would appear as duplicates of the Riemannian Stars that we already saw. But as far back as history records, no one has ever reported this. There should be two Stonehenges marking the movement of the stars. One would be the Euclidian Stonehenge and the other would be the Riemannian Stonehenge. Since the sun is a star, and since light passes through Riemannian Space, we should see two suns separated by about eight and a half minutes of earth rotation and revolution. (The sun is about eight and a half light-minutes away.) Go outside and look at your shadow on a sunny day. If light passes through Riemannian Space, then you should have two shadows. One shadow would be caused by the light of the sun that passed through Euclidian Space and the other shadow would be caused by the light that passed through Riemannian Space. Do you expect to see two shadows? You should if you're a Young-Earth Creationist who uses Riemannian Space as a way of defending the Young-Earth Theory. I could go on, but again we have people claiming that God makes things appear to be doing what they aren't doing. God again becomes deceptive. Dear Christian, we should think long and hard about what we preach before we preach it. If it impugns God's character, we should do what Job did when God confronted him with his ignorance and arrogance.

Job 40:3-5 "Then Job answered the LORD: *{4}* 'I am unworthy—how can I reply to you? I put my hand over my mouth. *{5}* I spoke once, but I have no answer— twice, but I will say no more.'" (NIV)

Why Not Flood-Geology?

The arguments for and against Flood-Geology are unending. Hundreds and hundreds of books, thousands and thousands of articles, and probably millions of pages of material have been written attacking and defending Flood-Geology. I cannot possibly cover more than a few basic reasons why I think Traditional-Geology disproves Flood-Geology. What is Traditional-Geology? Traditional-Geology is what most accredited, university-trained geologists would accept as scientific. Traditional-Geology is pretty much accepted by every geologist who isn't a Flood-Geologist. Traditional-Geologists include more than unbelievers too. There are plenty of Christian geologists who reject Flood-Geology. That's why there has been so much material written against it. If I had to guess, I would say there are many more Christian Traditional-Geologists than Christian Flood-Geologists, but that's just a guess. Anyway, PLEASE note very carefully that I did not say that Traditional-Geology disproves The Flood. I said it disproves Flood-Geology. Traditional-Geology disproves Flood-Geology but not The Flood because Flood-Geology has very little to do with The Flood. Instead, it is the creation of some very fertile minds using very few facts. (Dr. Dill's Gardening Tip #1: It doesn't matter how much fertilizer you add; if you don't have seeds, you won't get a crop.) Flood-Geology has very few "seeds" but lots of "fertilizer." It is largely based on false interpretations of Scripture and distortions of science. Since I don't want to get deluged by all the arguments, I won't go into all the details. I only want to point out a few glaring problems with Flood-Geology.

Specimen Ridge

Specimen Ridge is a large area in Yellowstone National Park. This area has an interesting arrangement of geological strata. There are about fifty layers that alternate between forest material and volcanic ash. Traditional-Geology explains it this way: Yellowstone was a volcanic area millions of years ago. A lush forest grew there, but suddenly a volcano erupted. When that happened, the forest was destroyed and was covered in volcanic ash. Over time a new forest grew. Years passed and another volcanic eruption destroyed and covered the new forest. The forest grew again and later another volcanic eruption covered the forest. The Traditional-Geologist tells us that several million years ago this section of Yellowstone experienced many, many such volcanic eruptions. Layer after layer of forest material and volcanic ash were formed. Today, millions of years later, tourists can visit this area to see the evidence for themselves. Of course if those tourists were Flood-Geologists, they couldn't accept this interpretation of the evidence. Flood-Geology says that this couldn't have happened millions of years ago. It couldn't be the results of new forests growing up after volcanic eruptions. That would take too long. The earth isn't old enough for that to happen. Instead, these layers must have been deposited by the Great Flood. The flood waters rose and brought in vast amounts of forest material from surrounding areas. The Great Flood was so catastrophic that it uprooted trees, entire forests in fact, and transported them for miles. The waters then suddenly calmed over the area of Specimen Ridge and the forest material settled out. Then the water gently receded so as not to wash away the newly deposited forest material. Once the water receded, a volcano erupted and blanketed everything in ash. Then the volcano stopped and the tumultuous waters rose again. This brought in a new load of forest material ripped from another location miles away. Now, somehow this cataclysmic flood was strong enough to carry tons and tons and tons of forest material, yet so gentle that it didn't disturb the new layer of volcanic ash. The flood became calm again and the new forest material was deposited on top of that new volcanic ash. Then the waters gently receded again. Then a volcano erupted again. Then a new layer of volcanic ash covered the new layer of forest material. Then the flood waters became violent again and swept in a new forest ripped from a new location miles away and deposited that on top of the new layer of ash. Then the water receded again. Then a volcano erupted again, and so on, and so on until fifty layers were deposited. Over the course of about a year, the waters of the Great Flood rose twenty-five times, receded twenty-five times, and twenty-five volcanic eruptions occurred in between those times. This means that about every two weeks the waters rose, fell, and a volcano erupted.

Is the Flood-Geology interpretation scientific? I don't think so, but then I'm not a geologist. Traditional-Geologists laugh at the "science" of Flood-Geologists. Flood-Geologists laugh back. Personally, I don't see how such powerful waves of water could have deposited new material without ripping out the freshly-deposited material in the layer below it. I find it difficult to believe that the volcanic forces in the great depths below Specimen Ridge knew exactly when to become active and when to go dormant so that the layers would alternate. I find it hard to believe that twenty-five volcanic eruptions would occur in the same area in less than a year. I'm not a vulcanologist either, but it was my impression that once a volcano erupts, the pressure below it lessens and a dormant period follows. Usually this dormant period is quite long. There is a lot of Flood-Geology science that doesn't make sense to me, but again let me direct you to your Christian bookstore. Christian Traditional-Geologists and Christian Flood-Geologists have lots to say about it. The science of Traditional-Geologists seems much more reliable to me, but let's say Flood-Geologists prove their view is right. Let's assume they have the scientific evidence that the waters of The Great Flood did exactly what they say. Let's let them win the scientific argument! Now that we have the "scientific" evidence that The Flood rose and receded twenty-five times, let's compare that with the biblical evidence. Let's see if the biblical account matches the "scientific" account. What does the Bible say about the water levels?

Gen. 7:10-24 "And after the seven days the floodwaters came on the earth. *{11}* In the six hundredth year of Noah's life, on the seventeenth day of the second month—on that day all the springs of the great deep burst forth, and the floodgates of the heavens were opened. *{12}* And rain fell on the earth forty days and forty nights. *{13}* On that very day Noah and his sons, Shem, Ham and Japheth, together with his wife and the wives of his three sons, entered the ark. *{14}* They had with them every wild animal according to its kind, all livestock according to their kinds, every creature that moves along the ground according to its kind and every bird according to its kind, everything with wings. *{15}* Pairs of all creatures that have the breath of life in them came to Noah and entered the ark. *{16}* The animals going in were male and female of every living thing, as God had commanded Noah. Then the LORD shut him in. *{17}* For forty days **the flood kept coming** on the earth, and as the waters **increased** they lifted the ark high above the earth. *{18}* The waters **rose** and **increased** greatly on the earth, and the ark floated on the surface of the water. *{19}* They **rose** greatly on the earth, and all the high mountains under the entire heavens were covered. *{20}* The waters **rose** and covered the mountains to a depth of more than twenty feet. *{21}* Every living thing that moved on the earth perished—birds, livestock, wild animals, all the creatures that swarm over the earth, and all mankind. *{22}* Everything on dry land that had the breath of life in its nostrils died. *{23}* Every living thing on the face of the earth was wiped out; men and animals and the creatures that move along the ground and the birds of the air were wiped from the earth. Only Noah was left, and those with him in the ark. *{24}* The waters flooded the earth for a hundred and fifty days." (NIV)

For the first hundred and fifty days, the water rose; it kept coming; it increased. It didn't decrease. The Bible doesn't say the water rose and receded, rose and receded, rose and receded. "The flood kept coming on the earth." The waters didn't recede. The Bible doesn't say a thing about the waters going down during the first hundred and fifty days. The biblical account doesn't match the Young-Earth "scientific" account. Furthermore, Genesis 7:19-20 says that the waters covered even the highest mountains to a depth of more than twenty feet. Hmmm? Let's see. Specimen Ridge has an altitude of less than 7,500 feet. Mount Everest has an altitude of about 26,500 feet. That means that when Mount Everest was covered by twenty feet of water, Specimen Ridge was covered by 19,000 feet of water. For the top of Specimen Ridge to be repeatedly covered by dry volcanic ash, the water had to drop 19,000 feet. Then the water went up 19,000 feet, then dropped 19,000 feet, then went up 19,000 feet, etc. for twenty-five times. According to Flood-Geologist, twenty-five 19,000 foot waves flooded over Specimen Ridge without disturbing any of the newly deposited layers of sediment. Let's look now at what the Bible says about the water levels dropping.

Gen. 8:1-5 "But God remembered Noah and all the wild animals and the livestock that were with him in the ark, and he sent a wind over the earth, and the waters **receded**. *{2}* Now the springs of the deep and the floodgates of the heavens had been closed, and the rain had stopped falling from the sky. *{3}* The water **receded steadily** from the earth. At the end of the hundred and fifty days the water had **gone down**, *{4}* and on the seventeenth day of the seventh month the ark came to rest on the mountains of Ararat. *{5}* The waters **continued**

to recede until the tenth month, and on the first day of the tenth month the tops of the mountains became visible." (NIV)

The Bible is very clear here. The water didn't go up and down. It "receded steadily." It "continued to recede" until the mountains were finally uncovered. As you read the chapter, nothing hints that the water was going up and down, up and down. The tops of the mountains didn't become visible until the first day of the tenth month. If the Bible is true, the mountains had been covered for months. The most straightforward and clearest understanding of the biblical account of The Flood is that even Mt. Everest was covered by more than twenty feet of water until the first day of the tenth month. Specimen Ridge, 19,000 feet below this, surely wouldn't have been visible until much later. Flood-Geologists say this isn't true. Specimen Ridge was covered, then uncovered, then covered, then uncovered for twenty-five times. If Flood-Geologists are correct, then the tops of thousands of other mountains would have been visible long before the first day of the tenth month. If the top six inches of Specimen Ridge were uncovered, 19,000 feet of Mount Everest would have been visible. I find it strange that God seemed to overlook Mount Everest being visible so many times before the first day of the tenth month. He also didn't see the other mountains. Every time the water receded enough for Specimen Ridge to be visible, every mountain taller than 7,500 feet would have been visible. Did God not see them? Did He forget to tell Moses what really happened? If what Flood-Geologists say is true, then the Bible gives us a false account of what really happened during The Flood. If what Flood-Geologists say about Specimen Ridge is true, then God didn't tell us the truth about The Flood. What He told us doesn't match with Flood-Geology, and it isn't a matter of Him merely leaving out some details. I'm sure Flood-Geologists will say that God simply didn't tell

Moses about the water going up and down, but God didn't
omit telling Moses what the water was doing. He said the
water, "kept coming on the earth." Three times He said it
rose, and then He said it kept rising. If the water had actually
risen and fallen, He could not have said what He said. That
would not be an omission. That would be a lie. In the same
fashion, God couldn't have said the water "receded steadily"
and "continued to recede" if it was undergoing increases and
decreases in depth. Again, that would not be an omission.
That would be a lie. **Young-Earth Flood-Geologists put
their time, energy, and emotion into a theory that winds
up proving God is a liar if they are right.** Oh, how the
atheists must be howling with joy.

Green River Varves

If you thought the waters of the Great Flood were magic
over Specimen Ridge, wait until you learn what they did
along the Green River in the western United States. The
Green River basin includes parts of Colorado, Utah, and
Wyoming. First, let me tell you about varves. Varves are
annual sediment deposits on the bottom of lakes, ponds,
or other bodies of still or slow-moving water. Varves are
formed when different kinds of waterborne particles settle to
the bottom. One of the most common types of varves is the
clay/pollen couplet. Rivers and streams carry fine clay particles in suspension. These clay particles settle to the bottom
in places where the water becomes still. During spring and
summer, the water also carries large amounts of pollen from
trees and plants. This also settles to the bottom where the
water becomes still. There are places along the Green River
basin where scientists have examined varves for decades.
What they have observed is that the bottom of the river basin
grows a new varve every year. There is a layer of clay silt
and there is a layer of pollen. They have been able to show

that the varves in the Green River are due to annual sedimentation of clay silt and pollen. Young-Earth creationists agree to a certain point. They accept the annual sedimentation explanation only back to the Great Flood. That means only the top 4,000 or so varves were caused by clay/pollen sedimentation. The varves seen in the river basin strata below that were caused by the Great Flood. Just as in the case of Specimen Ridge, the waters moved up and down. In this case, instead of bringing in uprooted forests, they brought in alternating deposits of pollen and clay silt. Now, I don't know much about hydraulics, the science of fluids, but I bet it would be pretty difficult to figure out how cataclysmic, raging waters could deposit so many microscopic particles so gently over hundreds of miles of river in less than a year. It would be especially difficult to explain if the area was under thousands of feet of water for most of that year. The biggest difficulty appears when you learn how many varves there are in the Green River Basin. There are places that have twenty million varves. Traditional-Geologists say the river has been there for millions of years. Young-Earth Flood-Geologists disagree. They have three explanations. The first, I have already mentioned. The Great Flood did it. This time, instead of twenty-five ebbs and flows of the Great Flood like at Specimen Ridge, there had to be twenty million ebbs and flows in one year. Does this make sense? There are about 31 million seconds in a year. Apparently they believe a wall of water came rushing down the Green River carrying clay silt. This was followed 0.78 seconds later by a wall of water carrying pollen. This was followed 0.78 seconds later by a wall of water carrying clay silt. This was followed 0.78 seconds later by a wall of water carrying pollen. This was followed 0.78 seconds later by a wall of water carrying clay silt. This went on for a year, and remember, it had to be even faster since it was completely underwater most of the time. This doesn't seem very scientific to me. Their second explanation is that God placed all those twenty million varves in place in order to give the Green River the appearance of being millions of years old. We've already talked about the concept of Apparent Age. This doesn't seem very biblical to me. Their last explanation is that these varves are not due to sedimentation of clay and pollen. Instead they are due to some chemical process. This is playing the "What-If Game." I'll talk more about the "What-If Game" later. For now let's examine this explanation. They have no evidence to defend it. There is plenty of evidence that these layers were caused by sedimentation of clay particles and pollen particles. Scientists have directly observed this for nearly a century. Microscopic examination of the layers reveals clay particles and pollen particles. True, there is a point at which the years have decayed the pollen so much that the pollen particles aren't intact. It is at this point that Young-Earthers say the varves were the results of some chemical reaction. Chemical analyses of the layers, however, still prove they are clay and pollen deposits. There is no indication they were caused by chemical reactions. There is nothing indicating there was a shift from a series of chemical reactions to a series of physical sedimentations. Claiming that it was a chemical reaction does not prove it was a chemical reaction. They insist they are right just because we can't prove they are wrong. This is neither scientific or God honoring. I could just as easily claim that space aliens came to earth a thousand years ago, dug the entire Green River Basin, deposited twenty million varves of clay silt and pollen, and then zapped the entire human race with a memory-erasing beam so that we wouldn't know they did it. You can't disprove that explanation either, but I'm not going to win a Nobel Prize with it. I don't think Flood-Geologist will win Nobel Prizes with their explanations either.

Coral Reefs

The waters of the Great Flood performed acts of magic by depositing all the layers of sediment at Specimen Ridge in between those volcanic eruptions. The waters of the Great Flood performed even greater magic by depositing all those varves along the Green River Basin. Yet, those feats were nothing compared to how The Flood constructed the world's largest coral reefs according to some Young-Earth creationists. Coral is actually a marine animal. Rather than swimming around its whole life looking for food, it likes to settle down, build a little calcium carbonate house, and wait for food to come to it. (Kind of like calling out for pizza.) Coral also likes to stay close to family. New coral will attach itself to older coral. This way it uses part of the old coral's house as part of its house. This saves energy for the coral. In time, as newer coral builds on previous layers of coral, a coral reef forms. Traditional-Geologists tell us it takes thousands and thousands of years to form coral reefs. Traditional-Geologists quote Traditional Biologists who say that the fastest a coral reef can grow is about one-half inch per year. This means that a mile-wide coral reef would have needed over 100,000 years to become that large. There are active coral reefs today that appear to be even older. This is the Traditional-Geologist/Biologist opinion. Flood-Geologists tell us that these coral reefs didn't grow on site. Instead, they were formed when the Great Flood ripped up smaller coral reefs all across the oceans and deposited them in these locations. Those coral reefs only APPEAR to be 100,000 years old. Flood-Geologists get an "F" in biology when they say this because coral is not the same as clay silt or grains of pollen. Those kinds of particles are deposited just willy-nilly by water, but this isn't true of coral. When coral polyps build their houses, they build them in a vertical orientation. They have a top to their house and they have a bottom to their house. They also have a different type of construction if they build their house on the seaward side of the reef as opposed to the landward side of the reef. Since they attach themselves to previously built coral houses, their walls fit the contours of the walls of the corals next to them. This means that if you took a chunk of coral reef and examined it under the microscope, you could tell which way was up, which way was down, which way was in, and which way was out. Then, like a giant jigsaw puzzle, you could fit individual corals together by the contours of their shells. If Flood-Geologists are telling us the truth, then the Great Flood ripped apart thousands of coral reefs and transported billions of individual coral units over hundreds or even thousands of miles. Then the waters of The Flood reassembled each coral unit in the correct up/down, in/out orientation. The waters also knew which coral to attach to which so that the contours of their walls again matched the very same neighbors they had been attached to before being transported. Magic water indeed!

This scientific problem doesn't bother the determined Young-Earth Flood-Geologist. His response is, "But what if God had done this? (The "What-If Game.") Couldn't God transport billions of coral units over thousand of miles and reassemble them exactly as they had been before? Would this be too difficult for God? Some Old-Earth creationists dismiss this question from Young-Earth creationists because such an act of God seems to make no sense. They ask, "Why would God do such a thing?" To which the Young-Earth Creationist responds by saying, "We don't know why, but we also don't know why God does a lot of other things. We believe on the basis of faith." This answer is intended to make it look as if Old-Earth creationists have no faith. The real answer, of course, is that God most certainly could have done this. God keeps track of the 10^{80} fundamental particles of matter in the universe every second. A few billion coral units wouldn't be a problem. It's not a question of WHY God would do this.

It's not a question of COULD God do this. The question is, "DID God do this?" On what basis do Young-Earth creationists believe that God did this to coral reefs? Do they have any scientific evidence for this? NO! Do they have a passage of Scripture that says God did this to coral reefs? NO! They have a biblical interpretation of creation that needs this kind of physical interpretation of creation. Yes, they believe on the basis of faith, but it is not faith in the Bible, it is faith in AN INTERPRETATION of the Bible.

My Sentiments on Sediments

The geological record reveals a distinct pattern of appearance of the various fossils. This is the basis for the Fossil Index System. Now, while the evolutionary interpretation of the geological strata is incorrect, the raw data is still valid. **Specific fossils and fossil types do appear in a fairly set order.** Smaller, simpler things do show up before the larger, more complex things. Why is this so? What could have caused this? Day-Age creationists say that God created things in that order, but as we have seen, that order doesn't coincide with the Genesis Creation Account. Young-Earth creationists say this was the result of sedimentation during The Great Flood. The main reason fossils appear where they do, they say, is because smaller organisms would have been more easily washed away by flood waters. Larger organisms such as mammals tend to be more buoyant. Size and density are the reasons all the drowned organisms settled out in the sediment the way they did. This explanation fails the scrutiny of logic. First of all, not all fossils are in water-borne strata. Gobs (not a scientific term) of fossils are found in dry, wind-borne sediment. I know God used The Flood to cleanse the world, but unless He followed it with a "spin-dry" period, I don't see how The Flood could create dry, wind-borne sediment with fossils buried in it. Second, size and density make no difference. Modern snails are the same size and density as ancient snails, but modern snails aren't found in the same layers. Turkey-sized dinosaurs are the same size as turkey-sized turkeys, but turkey fossils are never found alongside dinosaur fossils. If size, density, and buoyancy are the factors that determine where fossils appear in the geological strata, then modern fish should be right next to primordial fish, modern plants ought to be in the same strata as primitive plants, and modern insects should be the fossil neighbors of ancient insects. Young-Earth creationists counter this by saying we need to consider the mobility factor. Larger, more complex organisms are more mobile and could have run to higher ground when The Flood hit. Their fossils would appear later in the strata. This seems reasonable at first glance, but it too is illogical. Many dinosaurs could run faster than many mammals. Modern snails couldn't outrun a Tyrannosaurus. So why are Tyrannosaurus fossils always deeper than modern snails? Apple trees have the same mobility as ancient Bryophytes. Where are apple tree fossils and ancient Bryophyte fossils found in the same strata? Modern sleek, shiny, scaly fish have no more ability to run to the mountain tops than the primitive squishy, slimy, fishy-things that supposedly swam in the same lakes and same oceans at the same time. Why don't their fossils appear in the same geological strata? A prairie dog in Kansas had less chance of reaching the mountains than a Pterodactyl. So why don't we find prairie dog fossils in strata deeper than Pterodactyl fossils? Speaking of running dinosaurs, why have paleontologists found dinosaur nests[2] in between fossil-bearing strata? Some of these nests have eggs with fully developed baby dinosaurs inside. Since these nests are on top of fossil-bearing strata supposedly deposited by The Flood, these "fleeing" dinosaurs had to tread water until the waters receded. Then they took the time to mate, build nests, and lay eggs. They didn't seem in much of a hurry to get to the mountains. There seemed to be enough time for the eggs to incubate and the babies to develop before being covered by more fossil-bearing sediment when the water rose again. Nothing in the Bible indicates that animals did this during The Flood.

Nevertheless, if you are familiar with the arguments for and against Flood-Geology, then you know I would be lying if I said there were no examples of strata that seem out of place. There are many examples of younger fossils found beneath older fossils, and the arguments about how this happened are unending. Regardless of the explanations, these out-of-place fossils are still geological exceptions. The vast majority of fossils fit in very well with Traditional-Geology. They fit so well that almost any geologist can examine a fossil specimen and tell you where it fits in the geological scheme, and they would be right most of the time. This was why the Fossil Index System seemed like a great defense for evolution. Almost all fossils fit into the Traditional Geological position. When they are out of place, the explanations of the Traditional-Geologists (overthrusts, earthquakes, landslides) seem to me to be more scientific. Remember, I didn't always believe this. I once accepted what Young-Earth creationists told me, but over the years I have seen too many credibility problems. Too many Young-Earth arguments have proven to be bogus. This is usually due to lack of data or misinformation, but I have seen some outright lies and distortions. Many of the Young-Earth arguments I once used against evolution, I have had to retract because I discovered they weren't true. Let me give you some examples.

Carbon 14 Errors

You have probably heard about Carbon Dating. If you are not familiar with it, let me explain. Normal carbon is carbon 12. This is written as ^{12}C. Radioactive carbon is carbon 14. This is written as ^{14}C. ^{14}C is made from normal nitrogen, ^{14}N, when cosmic rays strike atoms in the upper atmosphere. It is assumed that the ratio of ^{12}C to ^{14}C is a constant. Plants take in carbon in the form of carbon dioxide in the air. As long as a plant is living, its $^{12}C/^{14}C$ ratio will be the same as the $^{12}C/^{14}C$ ratio in the air. When the plant dies, it no longer takes in new carbon. ^{14}C is unstable and decays over time. It has a half-life of about 5,700 years. This means that every 5,700 years, one-half of the ^{14}C in the dead plant material will decay back to nitrogen. (If you want to impress your friends, tell them this happens when it emits an electron and an anti-neutrino... they'll think you're smart.) By measuring the $^{12}C/^{14}C$ ratio in the dead plant material, scientists can calculate how old the plant material is. Since animals eat plants directly or indirectly, animals have the same $^{12}C/^{14}C$ ratio as plants. Again, as long as they are alive and eating, they will have the same $^{12}C/^{14}C$ ratio. When animals die, they no longer take in new carbon and the same radioactive decay process occurs. Carbon Dating is therefore used to measure the age of material that was once living. It can't measure the age of rocks and minerals. In addition, after eight or nine half-lives, the amount of ^{14}C is too small to detect. This limits Carbon Dating to 50,000 years or so. Now, there are a lot of assumptions made when using Carbon Dating. Young-Earth creationists have long accused Old-Earth creationists of making wildly incorrect assumptions for Carbon Dating. I don't think they have, but I'm sure you have your own opinion. The real problem for me was when I learned how some Young-Earth creationists "invalidate" the Carbon Dating System by using deceptive information... dirty tricks. (NOTE: I SAID "SOME," NOT "ALL.") There are some Young-Earth creationists who have taken tissue from living or freshly killed organisms and have had it Carbon Dated as being several thousand years old. They offer this as proof that Carbon Dating is invalid. Many people believe them. I did, but then I discovered what really happened. I also discovered that they knew what really happened, but they didn't share it with their audience. They hid the truth. The $^{12}C/^{14}C$ ratio in any organism is going to be the same ratio as in its food. Since plants take in carbon dioxide from the air, plants have the same ratio as in the air. Animals will have the same ratio ONLY if they consume new plant material. If you took 5,000 year old organic debris and fed it to organisms that feed on organic debris, then you would have organisms with a $^{12}C/^{14}C$ ratio that would be the same as in the 5,000 year old organic debris. They would look 5,000 years old according to Carbon Dating. In other words, **feed it 5,000 year old carbon and it will look 5,000 years old by Carbon Dating.** If you took these organisms and fed them to higher organisms, then those higher organisms would also have a Carbon Date of 5,000 years. Any biologic niche that is contaminated by "old carbon" will produce organisms that appear old according to Carbon Dating. This is what these Young-Earth creationists do. They find a place with "old carbon" to collect their samples. For instance, they find a river that has cut a channel or valley down through several geological layers. They go to a layer that is 5,000 years old. They take out living organisms that have been feeding on the 5,000 year-old organic debris in that layer. Then they use Carbon Dating to show how these living organisms have been "dead for 5,000 years." This deceptive technique is used to convince people that Carbon Dating is wrong. This is their "proof" that the earth is young. Ironically, they verify the Carbon Dating system. They use organisms made of 5,000 years old carbon, and Carbon Dating shows they are made of 5,000 year old carbon. The carbon dating is correct, but they don't tell us that! Instead, they tell us that Carbon Dating is erroneous, and they mock Old-Earth creationists for believing the earth is old. They make it look as if Old-Earth Christians are wrong. Is this something Christians should do? Where does the Bible say we can use lies and distortions to defend the Bible? If they intentionally lie, they bring shame to all Christians and to Christ. If they do this out of ignorance, then it's a little more excusable. However, they should know better. Unfortunately you can find Christian books, Christian magazines, Christian newsletters, Christian newspapers, Christian Internet Web Sites, Christian videos, and Christian radio programs that still use this dirty trick to fool us. **What are atheists to think about Christ when Christians deliberately lie?** This upsets me. I hope it upsets you too.

Japanese Plesiosaur

Young-Earth creationists believe that dinosaurs and humans were contemporaries 6,000 years ago. Old-Earth creationists disagree. They say there is no evidence for such a claim. Young-Earth creationists argue that they have proof for their belief. In 1977 fishermen on a Japanese fishing trawler hauled in a gigantic rotting carcass off the coast of New Zealand. The fishermen didn't know what it was and immediately speculated that it was a Plesiosaur, a marine dinosaur (reptile) that lived 100 to 200 million years ago. Young-Earth creationists were ecstatic. Here was a dinosaur that wasn't millions of years old. Young-Earth creationists immediately began using this discovery as a way to "prove" their theory was true. Some used this discovery as a way of ridiculing Old-Earth creationists. Here again, Mr. Deception raised his ugly head. Within a few weeks of the discovery, scientific tests indicated that the cartilage in the carcass came from a shark. It wasn't a reptile. Within a few months, amino acid analyses of its proteins proved that the carcass was that of a Basking Shark (*Cetorhinus maximus*). Basking Shark experts looked at the photographs of the rotting carcass and affirmed that it was a rotting Basking Shark. Some Young-Earth creationists still claim this was a Plesiosaur.

Paluxy River Footprints

After I learned that the Theory of Evolution was a lie, I began searching for answers about creation. I didn't know what to believe, but as soon as I saw photographs of human footprints in the same geological strata as dinosaur footprints, I became a Young-Earther. This was in 1972 and I was young and easily impressed. I consumed volumes of Young-Earth books and pamphlets to learn more about these footprints in the riverbed of the Paluxy River near Glen Rose, Texas. These weren't ordinary footprints. These were footprints of dinosaurs and humans in the same limestone. This seemed like absolute proof that men and dinosaurs lived together 6,000 years ago. The photographs were clear. It appeared that the dinosaur footprints and the human footprints were real. For years, Young-Earth creationists wrote article after article and book after book about how this "proved" they were right. Old-Earth creationists refuted their evidence, but that made the Young-Earth creationists more determined than ever to show the world their "proof." Then something very bad happened to their "proof." It was shown to be incorrect, and it was shown to be incorrect by some of the same Young-Earth creationists who had once believed it. It was eventually shown that the evidence was a combination of wild imaginations, hoaxes, poor data, misunderstandings, ignorance, and just a little bit of longing for fame and fortune. Yes, there was money in them-there footprints. The scientific evidence was wrong and I salute the integrity of the Young-Earthers who have rejected it. However, there are still Young-Earth creationists who peddle this kind of snake oil for personal gain and fame.

Magnetic Field Decay

The earth has a magnetic field around it. This is good. It helps shield the earth from charged particles emitted from the sun. It allows us to see the Northern and Southern Lights. It makes compass navigation possible. These things have nothing to do with creation, but there is one aspect of the earth's magnetic field that Young-Earthers use to defend their position. The magnetic field of the earth is decaying; it is getting weaker. Scientists have been able to determine the rate of decay and it seems to create a problem for Old-Earthers. If you know the rate of decay, then you can calculate the strength of the field in the past. This is what Young-Earthers have done, and initially, I was impressed by their claim. If you extrapolate backward in time even a few thousand years, you can calculate that the earth had a tremendously stronger magnetic field. If you go back millions of years, then the magnetic field of the earth would have been so great that it would have melted the crust of the earth. In the same fashion that magnetic-induction stovetops heat iron cookware, the magnetic field of the earth would have heated the iron in the crust to a point that it would have been molten. Obviously a molten earth would not be good for life. Even if dinosaurs had compasses they wouldn't be able to navigate very well on a molten surface, and they certainly wouldn't have enjoyed the Northern Lights. Therefore, the earth cannot be millions of years old. This sounds very scientific and very believable, and it is still being used to convince people that the earth is only a few thousand years old. Yet, like so many other Young-Earth scientific claims, there is more to the story than they tell. Yes, the magnetic field is getting weaker, but it hasn't always done so. There were times when it got stronger. The magnetic field is constantly changing. In fact, scientists have discovered that the magnetic field has reversed its polarity numerous times in the past. It gets weaker, then it changes polarity, then it gets stronger in the opposite direction, then it gets weaker, then it changes polarity. It cycles back and forth in its polarity. The needle on your compass does not point to the geographic North Pole, it points to the magnetic North Pole. The magnetic pole moves around. If you were alive hundreds of thousands of years ago, the needle of your compass would have still pointed to the magnetic "north" pole, but that would have been closer to the geographic South Pole. Still, many Young-Earthers use this argument to "prove" their point. This is as scientific as if I measured the temperature increase during the first few hours of the morning and then used this rate of increase to extrapolate ahead. If the temperature rose ten degrees in the first four hours after sunrise, I could then "prove" that within a week, the earth would be over four hundred degrees. Extrapolations don't work for cyclic phenomena.

Mississippi River Delta

I once believed the Young-Earth argument that the Mississippi River was less than five thousand years old. This was based on another extrapolation. The Mississippi River carries millions of tons of silt into the Gulf of Mexico every year. Therefore, the Mississippi River delta is growing larger. Scientists have been able to measure its rate of increase. By looking at the total area of the delta and dividing that area by the rate of increase, it is possible to show that it would take only 5,000 years to create the delta. This sounds very scientific, but like many Young-Earth arguments, it is based on incorrect assumptions. The delta hasn't always grown at the same rate. In fact, due to compaction, erosion, hurricanes, and tropical storms, the delta can even be washed away. It has actually gotten smaller at times. Extrapolations don't work here because we aren't dealing with a constant rate of growth. Now, if anyone has a right to use extrapolations dealing with the Mississippi River delta, it's Old-Earth Geologists. You see, Young-Earth Geologists weren't telling the whole truth. The delta is not only gaining in area, it's gaining in depth. This is a three-dimensional problem, but Young-Earthers have applied only two-dimensional explanations. So much sediment has been deposited that its weight has actually pushed the earth's crust down. How far down has the crust beneath the Mississippi River delta been pushed? Oh, about seven miles. What is most interesting is that the crust below the delta is sedimentary strata supposedly laid down by The Flood. It has geological strata loaded with fossils. A seven-mile pile of river sediment is sitting on top of fossil-laden strata. This creates some problems. First, when you use extrapolations to determine the age of the delta, you discover that it would take millions of years to deposit that much volume (not area) of river sediment. In addition, how could a Flood that was five miles deep (fifteen cubits above Mt. Everest) carry a seven mile thick deposit? Finally, if the Mississippi River was formed as The Flood receded, how could that much river sediment come to rest on top of the surface of the strata just created by The Flood? If seven miles of sediment was suddenly dumped on top of the strata, there should be a seven-mile high mountain at the end of the Mississippi River.

Young-Earthers explain what "really" happened. It just so happened that the horrendous earthquakes and tectonic forces that accompanied The Flood caused the earth to sink seven miles at the mouth of the Mississippi. A seven-mile deep hole suddenly formed. Then a seven-mile pile of river sediment quickly fell into that seven-mile deep hole. It was lucky the hole in the ocean floor just happened to be there. Otherwise, a seven-mile high wall of sediment would have been deposited at the mouth of the Mississippi River. If that had happened, much of North America would be under water today. It was lucky that those massive earthquakes were able to create a hole in just the right spot. Even luckier is the fact that other massive earthquakes just happened to create other massive holes in the ocean floor at the mouths of every other major river. The Nile, the Amazon, the Yangtze, and others have deep depressions in the strata at their mouths, and their holes are all filled up with river sediment too. What a coincidence that all those deep depressions were formed in just the right spots. Now, let's ask two key questions. First, is this Young-Earth explanation scientific? No, they have no scientific evidence that the deltas of the world's rivers were formed in this fashion. Second, is this Young-Earth explanation biblical? No, and this is going to shock some people. **There is no biblical statement that says earthquakes accompanied Noah's Flood.**

Gen. 7:11 "In the six hundredth year of Noah's life, on the seventeenth day of the second month—on that day all the springs of the great deep burst forth, and the floodgates of the heavens were opened." (NIV)

All it says is that the fountains of the deep burst forth. All the undersea springs suddenly released water, BUT IT DOESN'T SAY THERE WERE EARTHQUAKES. Adding earthquakes to the formula is a biblical interpretation, not a biblical declaration. The Bible doesn't say that earthquakes accompanied The Flood. The Bible doesn't say the earth's crust broke along the mid-Atlantic seabed as some Young-Earth creationists claim. The Bible doesn't say the continents were forced apart. The Bible doesn't say the continents were slammed together. The Bible doesn't say there were massive volcanic eruptions during The Flood. The Bible doesn't say that great holes formed in the ocean floor at the mouths of where the major rivers would later flow. The Bible doesn't say that Noah's Flood washed mountains into the sea. The Bible doesn't tell us that great tectonic forces created new mountains after The Flood. The Bible doesn't describe any of these events. All these events are creations in the minds of Young-Earth creationists. God didn't tell Moses about these colossal geological events. Did He forget? Here again, we have Young-Earth creationists defending biblical interpretations, not biblical statements. **They are not defending what the Bible says; they are defending what THEY WANT the Bible to say.** They have a twisted biblical interpretation of creation that needs this kind of twisted scientific interpretation of creation to defend it.

I could list many more holes in Flood-Geology. I won't because that isn't the purpose of this book. However, if you want to learn more, then I recommend Alan Hayward's book, *Creation and Evolution*[4]. He covers these arguments and others in much more detail. I have purchased multiple copies of Dr. Hayward's book over the years so that I can lend them to friends and not worry about them not being returned. It is a permanent part of my library. It should be part of your library too. Dr. Hayward does an excellent job at debunking Flood-Geology. For now, I think I have listed enough evidence to make my point clear. Young-Earth Theology is much worse than Day-Age Theology at explaining the geological strata.

If we view all this from the perspective of the Restoration Theory, we see how it fits. The geological strata record the events of the Pre-Adamic world, the first beginning. The six days of Genesis record the events of the restored world, the second beginning. The order of restoration doesn't match the order of creation because they were different events at different times. The order in which things were first created is not mentioned in Genesis 1:1. This means the geological strata shouldn't be expected to agree with the events described during the six days. The geological strata we see today are the results of pre-restoration geological forces, modified to some degree by the Great Flood and current geological forces. If this true, then some aspects of geology would indicate old ages, while other aspects of geology would indicate young ages. This is exactly what geology reveals if Young-Earth and Old-Earth Geologists are telling us the truth about what they discover. Both Young-Earthers and Old-Earthers do seem to have some valid arguments. There are some things that look young and there are some things that look old. When an Old-Earther proves that some piece of the puzzle is old, then he is looking at something from the first beginning. When a Young-Earther proves that some piece of the puzzle is young, then he is looking at something from the second beginning. **The Gap Theory removes the conflict between these two views.** Old-Earthers and Young-Earthers don't have to be enemies.

Chapter Six:

A Closer Look at the Bible

—⟋ⱳ⟍—

If I quit right here, then it is possible that some of you would be convinced that the Day-Age Theory and the Young-Earth Theory fail to answer some crucial scientific questions. They lack scientific evidence. On that basis, you might agree that the Gap Theory provides a better solution. Most of you, however, would probably not be convinced. I doubt I have changed the minds of many Day-Age and Young-Earth creationists. To be honest, if I were a diehard Day-Ager or a Young-Earther, I probably wouldn't be convinced either. The reason I wouldn't be convinced is not because the scientific evidence was lacking, but because I would feel a higher obligation to defend the Bible. As long as I believed the Bible truly taught that the universe was 6,000 years old, I would continue to defend the Young-Earth Theory. As long as I believed the Bible truly taught that the days were long ages, I would continue to defend the Day-Age Theory. What I really need to know is this: WHAT DOES THE BIBLE TRULY TEACH?

Is the earth young? Is it old? You'll have to decide for yourself, but I'd like to share my opinion with you. The Gap Theory suggests that both ages are valid because the earth had two different beginnings. The Gap Theory suggests that the earth had a beginning that may have been billions of years ago, and another beginning that may have been thousands of years ago. If the Bible reveals that earth had two different beginnings, then the Gap Theory is the most plausible theory. Is this what the Bible reveals? Let's start by looking at how several Hebrew scholars translate the first two verses of Genesis.

Genesis 1:1-2a

"In the beginning God created the heaven and the earth. And the earth was without form, and void; and darkness *was* upon the face of the deep" (KJV)

"In the beginning God created the heavens and the earth. The earth was without form, and void; and darkness *was* on the face of the deep" (NKJV)

"In the beginning God created the heavens and the earth. The earth was unformed and chaotic, and darkness lay upon the face of the deep" (BV)

"In the beginning God created the heavens and the earth. The earth was formless and empty, and darkness lay upon the face of the deep" (NBV)

"In the beginning God created the heavens and the earth. But the earth was empty and void, and darkness was over the face of the abyss" (Martin Luther Translation)[5]

"When God in the beginning formed the heaven and the earth, (then) the earth was waste and void, and darkness was upon the face of the deep" (Von Bohlem Translation)[6]

"In the beginning God **Elohim** created the *heaven* **heavens** and the earth. And the earth *was without form* **became waste**, and void; and darkness was upon the face of the *deep* **abyss**." (ERRB) (emphasis in original)

"When God set about to create the heaven and the earth, the world being then a formless waste, with darkness over the seas" (Speiser Translation)[7]

"In the beginning when God created the heavens and the earth, the earth was a formless void and darkness covered the face of the deep" (NRSV)

"In the beginning God created the heaven and the earth. But (then) the earth became waste, and darkness was upon the face of the deep" (Dillman Translation)[8]

"In the beginning of God's preparing the heavens and the earth, the earth hath existed waste and void, and darkness is on the face of the deep" (Robert Young Translation)[9]

"When God began to create the heavens and the earth, the earth was a desolate waste, with darkness covering the abyss" (S&G)

"In the beginning God created the sky and the earth. The earth was empty and had no form. Darkness covered the ocean" (EB)

"In the beginning which was not a beginning, in eternity past, ELOHIM (the Son) created out of nothing the entire universe, including Planet Earth. But the earth had become desolate and empty with darkness on the face of the raging waters" (Robert B. Thieme Translation)[10]

"In the beginning God created the heavens and the earth. And now, as far as the earth was concerned, it was waste and void, and darkness was upon the face of the deep" (Leupold Translation)[11]

"In the beginning God created the heavens and the earth. Now the earth was a formless void, there was darkness over the deep" (NJB)

"At the beginning of the creation of the heavens and the earth, the earth it was without form or life, and darkness was upon the face of the deep" (Cassuto Translation)[12]

"In the beginning God created the heaven and the earth. But the earth had become a ruin and desolation; and darkness was upon the face of the deep" (Arthur Custance Translation)[13]

"In the beginning God created the heavens and the earth. Now the earth was formless and empty, darkness was over the surface of the deep" (NIV)

"When God began to form the universe, the world was void and vacant, darkness lay on the abyss." (MB)

"In the beginning God created the heaven and the earth. The earth was unformed and empty; clouds covered the abyss" (English translation of the Crampon Translation)[14]

"In the beginning, God created the heaven and the earth, But the earth was unsightly and unfurnished, and darkness was over the deep" (Septuagint, Bagster Edition)[15]

"In the beginning God created the heavens and the earth. Now the earth had become waste and wild, and darkness was on the face of the roaring deep" (TEB)

"In the beginning Elohim created the heavens and the earth. And the earth came to be (the earth became) formless and empty, and darkness was on the face of the deep." (SCRIP)

"In the beginning God created the heavens and the earth. And the earth was formless and void, and darkness was over the surface of the deep" (NASB)

"In the beginning God created the heavens and the earth. And the earth was waste and void and darkness was upon the face of the deep" (RSV)

"Look, look, look. See God. See God create. See God create the heavens and the earth. Create, God, create! Oh, oh, oh, see the earth. The earth is without form, and void..." (The New Dill Nonstandard Translation)

Isn't it amazing how many different translations of Genesis 1:1-2 there are? I don't know what it says about mankind, but we can't even finish the first sentence of the first paragraph of the first chapter of the first book of the Bible without creating a controversy. There seems to be about as many translations as there are Hebrew scholars. I suppose it's only human nature for each Hebrew scholar to come out with the best translation ever. I've also noticed that nearly all of these scholars are the world's foremost authority in ancient Hebrew, but very few totally agree with how the others translate. They all tend to say things like, "Yes, it can be translated the way Dr. So-And-So puts it, but..." At this point they show you how wrong Dr. So-And-So is and how right they are by translating it the way they believe.

After reading a number of books dealing with the Genesis account of creation, I have made a couple of deductions about the translations of Genesis. Before I mention those deductions, however, I want to reveal my own authority and expertise in this area. I have none. I cannot read Hebrew. I don't know a *Niphal* from a *Hiphil* or a *Piel* from a porcupine. I can, however, read the comments of those who do know *Piels* from porcupines. If you are a Hebrew scholar and wish to criticize my opinion, please feel free to do so. If you aren't a Hebrew scholar and you hear experts criticizing me on the grounds that I'm not a scholar, I ask you to remember one thing. My opinion on the translation of the creation accounts is not MY opinion. It comes from a number of Hebrew scholars who are highly qualified and respected in this area. My role in this is similar to the little boy who knew nothing about chess, but challenged the world's top two chessmasters to play him simultaneously. He claimed he would beat at least one of them, or at worst, tie them both. His only request was that he be allowed to move second against one of them and move first against the other. To everyone's surprise he played brilliantly. He lost to one of the chessmasters, but as he had predicted, he beat the other. When the losing chessmaster asked him how he could play such superb chess while knowing nothing about the game, he revealed his secret. He merely repeated the moves of the first player against the second, and responded against the first player with how the second player responded to those same moves. In reality he played one chessmaster against the other. That's what I'm doing when I enter the area of Hebrew translations. I quote the scholars and quote their quotes about other scholars' quotes. So when I tell you my opinion of how the Hebrew is supposed to be translated, it is not my opinion. It is the opinion of Hebrew scholars.

As I said, I've made two deductions about the translations of the Genesis account. My first deduction is that Hebrew must be a "rubber" language. Evidently, its rules of grammar and varied word definitions make it highly flexible. Meanings, seemingly, can be stretched to define almost anything. In truth, every one of the above listed translations (excluding mine, of course) can be defended by the rules of Hebrew grammar. Every one of those translations was made by knowledgeable and fully qualified Hebrew scholars. We non-scholars must tread lightly. We are at the mercy of the scholars, so I ask my readers to be wary. Don't let anyone tell you that even one of those translations of Genesis 1:1-2 is without grammatical support. He who makes such a statement is a liar. Every one of those translations has ardent supporters who are noted and well-educated Hebrew scholars. This doesn't mean they reveal the exact same information. While they are all **grammatically correct**, they cannot all be **informationally correct**. (I'll explain this later.) My second deduction is about Hebrew scholars. I'm sure this isn't true of all Hebrew scholars, but it probably describes the majority. I have concluded that each scholar translates the text based on what he believes the Bible says. This may not seem like a startling statement, but if you'll think about it for a minute or two, you'll see the significance. If a translator has a preconceived idea of what the text is suppose to say, then it shouldn't be a big surprise when he translates it the way it's suppose to be. I would venture to guess that most biblical Hebrew scholars studied the Bible during the process of becoming biblical Hebrew scholars competent enough to translate the Bible. The question then arises, "How did they get the knowledge it takes to become a biblical Hebrew scholar?" I suspect they studied biblical Hebrew in a seminary, a university, a Bible college, or some other establishment of higher education. This being the case, it seems only natural that scholars would tend to know what they've been taught. It wouldn't be strange if the majority of scholars trained in a Southern Baptist seminary, for instance, translated the text in a fashion similar to how Southern Baptist seminaries believe it is supposed to be translated. Likewise, if you were trained in a Presbyterian seminary, you might have a tendency to translate the Hebrew in a fashion similar to what Presbyterian seminaries teach. A young man studying to be a Catholic priest would probably translate the Hebrew in a way that fell into line with Catholic teachings. If you were taught Hebrew by scholars who believed the *King*

James Version was the only valid translation, then it would make sense that you'd translate Genesis in a way that would compare favorably with the *King James Version.*

You shouldn't be astonished to discover that the various Biblical creation theories center on what is historically believed by different churches, denominations, and other organizations. Many organizations hold strongly to their opinions and often require members and prospective members to believe and adhere to the very opinions they seek to defend. One would expect their translations and teachings to agree with their translations and teachings. I have not read many accounts of scholars agreeing with translations of the Bible that might disprove their own preconceived theories. (One notable exception was Gleason Archer. I'll tell you more about him in the next chapter.) There don't seem to be many "Frank Morisons" among the Hebrew scholars. In case you haven't heard of Frank Morison (That was his pen-name; his real name was Albert Henry Ross, 1881-1950), he was a British journalist who set out to use historical evidence to disprove the New Testament and the Gospel of Jesus Christ. However, when he was confronted with the historical facts, he realized his error, repented, and accepted Christ as his personal savior. He then used historical evidence to prove the truth about Christ. Now, I'm not accusing Hebrew scholars of being academically dishonest; just of being human. We all tend to think and believe what we were taught to think and believe, and we all tend to defend our thoughts and beliefs. We all tend to get upset when others challenge the things we believe. Nobody likes to have his opinions disproved or ridiculed. This is especially true for those of us with academic initials following our names. We don't like to admit that our interpretations could possibly be wrong. However, if we are Christians, then we should be more than willing to let truth prove or disprove our opinions. If we are wrong, then we should reject our false opinions and willingly accept the truth. Let's begin a detailed search for the truth about the creation as revealed in the Bible.

Seeking Informational Correctness

Genesis 1:1

"IN THE BEGINNING"

Some view this as the absolute beginning of everything, including time and space. There was no period of time prior to this because there was no such thing as time "before" this. Others consider this only the beginning of God's formative acts, but not the absolute beginning of absolutely everything. Speiser seems to indicate this in his translation:

"When God set about to create the heaven and the earth, the world being then a formless waste..."

"When" implies that time existed prior to the creation. "Being then a formless waste" implies that the earth already existed. Von Bohlem, Robert Young, Smith and Goodspeed, Moffatt, and *The New Revised Standard* translations seem to have similar opinions. They believe that Genesis describes a formation of the universe; not an actual creation.

"GOD CREATED"

At least everybody agrees on who the Creator was. It was God. The opinions vary on whether "created" means out of nothing or merely fashioned out of what was already there, but at least we give God the credit.

"THE HEAVEN"

Several translations say, "heaven," while others say, "heavens." Thieme lumps "heavens and earth" together as "the entire universe," while *The New Century Version* says, "the sky and the earth." The Hebrew word *SHAMAYIM*, translated "heaven," is a plural word, and should be rendered "heavens." However, I can't disprove the claim that it's in the plural form in order to emphasize the majesty of it all. The ancient Hebrews apparently pluralized words to indicate a greater emphasis. Among the commentators, there are those who feel that "heavens" include the sky, the universe, and the heaven where the angels abide. Others think this refers only to the atmosphere and space.

"AND THE EARTH"

"The earth" pretty much means "the earth" as far as I can tell from the scholars.

Genesis 1:2

Okay, now that we can't agree on Genesis 1:1, let's move on the Genesis 1:2 and see if we can continue the confusion. You might be wondering why there is so much concern over the relationship between Genesis 1:1 and Genesis 1:2. Here's where things start getting messy. Is Genesis 1:1 a title? Is it a summary? Is it part of the sequence of events or is there a break in the action? Does Genesis 1:2 follow immediately after Genesis 1:1? You might be asking yourself, "How does Genesis 1:1 fit with Genesis 1:2?" Well now, aren't you clever to ask that? That's exactly what I was going to talk about next. You'll love this! This is the stuff controversies are made of, so grab a handful of scathing comments and indignant rebuttals, and let's look at Genesis 1:2.

"AND"

Genesis 1:2 begins with the Hebrew word *WAW*. (Sometimes written as *VAV*) *WAW* happens to be one of those famous "Hebrew-rubber-words" with a variety of meanings. *WAW* is translated as "and," "but," "now," "then," "now then," "but then," "and then," and as other such linking-words. There is no end to the arguments over how *WAW* is supposed to be translated. Scholars tend to agree that *WAW* can be translated all these ways, but according to scholars, it is the context that often determines how it should be translated; and that's the problem! We have completed only one sentence and there is practically no context. Naturally, scholars translate it a certain way because they have already decided what the context is suppose to be.

Let me give you an example of how context causes *WAW* to be translated different ways. If I said, "The apple was ripe but it was bitter," then you would get the idea that being bitter isn't the normal condition of ripe apples. If I said, "The apple was ripe and it was sweet," then I would have conveyed to you that sweetness is something that ripe apples normally are. In English we have two separate words "but" and "and." In Hebrew, *WAW* would be used for both "and" and "but." You could say the apple was ripe *WAW* (but) bitter. You could also say the apple was ripe *WAW* (and) sweet. If you came across the Hebrew words *TAPPAWACH* (apple) *BASHAL* (ripe) *WAW* ("and" or "but") *MAR* (bitter), then you wouldn't necessarily know if *WAW* was to be translated "and" or "but" unless you knew how ripe apples were supposed to taste. If you didn't know what apples were, or if you had never tasted a ripe apple, then you might incorrectly translate it as, "The apple was ripe **and** it was bitter." You might think that ripe apples are normally bitter. If you knew that ripe apples are normally sweet, then you would translate it as, "The apple was ripe **but** it was bitter." Even though both ways of translating it are **grammatically correct**, you'd have to know something about ripe apples before you could translate this sentence in a way that would reveal the **correct information**. In this example, the "context" is the previously known fact that ripe apples are not supposed to be bitter. With this previous knowledge in mind, you would translate *TAPPAWACH BASHAL WAW MAR* as, "The apple was ripe BUT it was bitter." (Note to Hebrew scholars: Please forgive me for being so simplistic. I realize this isn't proper Hebrew grammar. I'm just trying to make a point on how the translation of *WAW* varies with context.)

Do you see how *WAW* can be translated "and" or "but," and be GRAMMATICALLY correct but INFORMATIONALLY incorrect? "And" implies that Genesis 1:1 is part of the sequence of events, while "but" implies that there is a break or a contrast in the action between Genesis 1:1 and Genesis 1:2. If you believe there is a break or a contrast between Genesis 1:1 and 1:2, then you are more likely to translate it "but." If you believe Genesis 1:1 flows into Genesis 1:2, then you are more likely to translate it "and." If you are a Hebrew scholar, you're not likely to translate it "but" if you believe Genesis 1:1 and 1:2 flow together in a rapid sequence of events. Likewise, if you're a Hebrew scholar who believes there is a gap between Genesis 1:1 and 1:2, then you probably aren't going to translate it "and."

We can correctly translate Genesis 1:2 grammatically, but we have to know what happened at the beginning before we can correctly translate it informationally. We have to know whether we need an "and" or a "but" to determine how *WAW* should be translated. The "context" in this situation is the previously known fact of whether or not there was a break in the creative works of God between Genesis 1:1 and 1:2. This is one of those situations where we have to know the truth before we know what is true.

WAW is also translated "now" in some of the translations. All three of these translations are grammatically correct, but each carries a little bit different information. As a
conjunction, "and" implies that Genesis 1:2 is a continuation of Genesis 1:1. As a disjunction, "but" implies that there
is some kind of break or contrast between Genesis 1:1 and
Genesis 1:2. "Now" merely calls attention to Genesis 1:2
without specifying a particular relationship to Genesis 1:1.
WAW, when translated "now" can also imply "now then."

Note also that a translation for *WAW* is completely left
out in some versions. That too is grammatically correct.
However, since it is in the original Hebrew, I don't think
WAW should be overlooked too lightly. The Holy Spirit
inspired Moses to use *WAW*, and I question the motives for
ignoring it. By starting a new sentence with a linking-word,
it forces the reader to think about the link between what the
sentences reveal. The first sentence reveals God's creative
acts and the second sentence reveals a desolate earth. There
must be some connection or relationship between God's creative acts and earth's desolate condition. By using *WAW*, I
think the Holy Spirit wanted to make us wonder about that
connection. Leaving *WAW* untranslated changes the information we obtain from these sentences. It's my opinion that
WAW in this position eliminates the possibility that Genesis
1:1 is a title. If Genesis 1:1 is a title, then the Bible actually
starts with the linking-word, *WAW*. If the Bible starts with a
linking-word, then to what is it linked? There is no previous
Scripture to which we could compare, contrast, or relate the
desolate condition of the earth. If *WAW* is left out, then the
description of creation begins in mid-sentence.

But *WAW* is correctly translated a number of ways. And
it is grammatically correct to ignore it in translation. Now if
someone tells you that it can't be translated "and," or "but,"
or "now," or that it can't be left out, then you have my per-
mission to laugh in their face. But in case they don't believe

you, have them look up every single "and," "but," and "now" in a concordance and check out the corresponding Hebrew words. And since there are over 19,000 "ands," "buts," and "nows" in the *King James Version* of the Old Testament, they should stay busy for awhile. And I know it's not quite proper to start an English sentence with a conjunction or a disjunction. But I wanted you to get a feel for how these words force you to compare the relationships between sentences. Now back to our study.

"THE EARTH"

Again, this is generally translated as "the earth," the little speck of dust in the universe inhabited by a race of creatures who inherently love controversy. In fact, if an issue isn't controversial enough, we seem to have a way of making it so. Sadly, there is even a controversy over what *ERETS* means in Genesis 1:2. Everyone believes that *ERETS* in Genesis 1:1 is the earth, but some insist that *ERETS* in Genesis 1:2 is the Land of Israel. You see, *ERETS* can be translated earth, ground, field, dirt, or land. They say it refers to "The Land," which is Israel. In this view, Genesis 1:3-31 describes God creating life, etc. in the land where Israel will someday be. Israel, therefore, is the center of all creation. You can probably guess who might favor this view. This translation stretches an already "rubber" word. The context seems to indicate that this is the same *ERETS* (earth) mentioned in Genesis 1:1, and I have yet to see a Hebrew scholar translate Genesis 1:1 as, "In the beginning, God created the heavens and Israel."

"WAS"

Starting with "was without form, and void," the controversies worsen. At this juncture brotherly love gets laid aside. Kid gloves are replaced by brass knuckles. It's not just the scholars' ability to translate Hebrew that's at stake. Entire churches, denominations, and schools have argued over this sentence. Reputations are on the line. There's too much at stake to let some idiot translate this in a fashion other than the way my church, my denomination, and my organization say it should be. Every knowledgeable scholar agrees that I'm right. If you don't believe it, just ask the knowledgeable scholars who agree with me. God wouldn't dare create the universe in a fashion that would disagree with what I believe. The Bible can't teach something contrary to what I... Oops, please excuse me. Momentarily I started thinking I was a scholar.

The biggest controversy of them all deals with how the Hebrew word *HAYAH* should be translated in this context. *HAYAH* is similar, but not equivalent to the English "to be." In its various forms it can be translated "be," "am," "are," "is," "was," "will be," "will become," "had been," "has been," "had become," "has become," "became," "come to pass," "came to pass," and a whole list of others. *HAYAH* is such a "Hebrew rubber-word" that if you dropped it on the floor, it would bounce around for a week. In fact, *The New American Standard Bible* has over 120 different translations for *HAYAH*. If you'll look back at the various translations you'll see that most of the time it's translated "was," but there are some exceptions. Custance translates it "had become." It is also translated "had become" by Rotherham (*The Emphasized Bible*) and by R. B. Thieme. Jahn (*The Exegeses Ready Research Bible*) translates *HAYAH* as "became," and Dillman does the same. *The Scriptures* translates it "came to be." Speiser translates it "being then," while Robert Young uses the words "hath existed." In more modern English, Young would rephrase it "has existed." Genesis 1:2 in the *King James Version* of the Bible says:

"And the earth **was** without form, and void; and darkness *was* upon the face of the deep. And the Spirit of God moved upon the face of the waters."

The word "was" is used twice in this verse, but there is a difference. The second time it is used, it is in italics. It says, "and darkness *was...*" not, "and darkness was..." This is important. In the *King James Version*, italics are used to indicate the words that are not in the original Hebrew or Greek, but have been added by the English translators to help the text make more sense to English readers. Continue reading and you will see numerous cases of "was" and numerous cases of "*was.*" Why the difference? Hebrew is one of those languages that doesn't need a form of "to be" in order to complete or connect a thought. In English I would say, "Linda Dill is my wife." In Hebrew it isn't necessary to add "is." It would be proper to say, "Linda Dill my wife." The word "is" really adds nothing to the thought. The first "was" in the Hebrew of Genesis 1:2 is there, but the second "*was*" isn't.

The second "*was*" is not in the Hebrew because it is not needed in Hebrew. (Or maybe I should say, "It not in the Hebrew because it not needed in Hebrew.") No *HAYAH* is needed in the second part of this sentence because Hebrew carries the same information whether it is there or not. So, when *HAYAH* is used, it is often for a reason other than just to complete or connect a thought. *Strong's Dictionary of the Hebrew Language*[16] defines *HAYAH* this way:

"H1961. hayah, haw-yaw'; a prim. root; to exist, i.e. be or become, come to pass (always emphatic, and not a mere copula or auxiliary)..."

Note that Strong says that *HAYAH* is not a mere copula; not a mere connecting word.

The New American Standard Exhaustive Concordance of the Bible[17] says this:

"H1961. hayah; a prim. root; to fall out, come to pass, become, be..."

While there are many correct translations of *HAYAH*, the major point of conflict centers around whether *HAYAH* should be translated as some form of "was" or some form of "became." Some creationists say the passage should be translated, "...the earth WAS without form, and void..." Others say it should be translated, "...the earth BECAME without form, and void..." You would think the grammatical form would decide how to translate it. Unfortunately, as soon as you read one scholar who tells you one thing, you'll find two other scholars who tell you something else. Whether you translate it "was" or "became" seems to depend on two things:

1) The context
2) Your preconceived ideas about the context.

How you translate *HAYAH* in Genesis 1:2 depends on what you want Genesis 1:2 to say. Of course, this depends a great deal on what you were previously taught about the creation. People have preconceived ideas about what Genesis is supposed to say, and they will often do anything or say anything to defend their opinions. A zealous opinion is a poor substitute for truth. I think you can see now what I meant when I said that while all the different translations may be grammatically correct, they are not all information-ally correct. All those translations of Genesis 1:1-2 ARE grammatically correct. Each translator worded it the way he did because he previously knew (or thought he knew) some-thing about the creation that supplied the context on which he could determine the best way to translate it. While each translation is grammatically correct, they don't all carry the same information. Look back at Thieme's translation for instance. He incorporates the concept of the space-time continuum by letting us know that "the beginning" really wasn't a beginning because it "began" in eternity past, not in time or space. This is a very clever translation, but to translate it this way, Thieme previously had to know that space and time did not exist before the creation. Only God existed before the creation. Thieme wasn't an eyewitness to creation, so how could he know that time didn't exist "before" the creation? How could he know that space didn't exist "before" the creation? He could attain this knowledge only by studying other parts of the Bible. Look at John 1:1-3.

John 1:1-3 "In the beginning was the Word, and the Word was with God, and the Word was God. {2} The same was in the beginning with God. {3} All things were made by him; and without him was not any thing made that was made." (KJV)

John clearly tells us that everything that exists was made by Jesus. If Jesus didn't make it, it never came into existence. Time exists; therefore time had to be made by Jesus. Space exists; therefore, space had to be made by Jesus. John lets us know that God preexists all things, including time and space. John let's us know that everything, which includes the space-time continuum, was made by God and did not exist eternally. There was no eternal matter and energy before the creation. Creation was a real creation, a creation out of nothing, and not merely a formation. With John's information in mind, we can understand how it would be informationally correct to translate Genesis 1:1 in such a way as Thieme did. Even time and space had their beginning at the beginning. It would be informationally wrong to translate Genesis 1:1 in such a way that might indicate matter and energy were co-eternal with God. Even though Genesis 1:1 can be translated that way grammatically, it would not harmonize with John 1:1-3 and the rest of the Bible. (Plus, it would not harmonize with science. Science has now proven that time and space had a real beginning.) It is important to understand that the key to translating Genesis 1 is found in what the rest of the Bible tells us about the creation. For this reason, I have to reject all translations of Genesis 1:1 and 1:2 that are grammatically correct, but cause the Bible to contradict itself.

"... WITHOUT FORM, AND VOID; AND DARKNESS *WAS* UPON THE FACE OF THE DEEP."

"... without form, and void; and darkness *was* upon the face of the deep" (KJV)

"... without form, and void; and darkness *was* on the face of the deep" (NKJV)

"... unformed and chaotic, and darkness lay upon the face of the deep" (BV)

"... formless and empty, and darkness lay upon the face of the deep" (NBV)

"... empty and void, and darkness was over the face of the abyss" (Martin Luther Translation)

"... waste and void, and darkness was upon the face of the deep" (Von Bohlem Translation)

"... a ruin and desolation; and darkness was upon the face of the deep" (Arthur Custance Translation)

"… waste and void and darkness was upon the face of the deep" (RSV)

"… unformed and empty; clouds covered the abyss" (English translation of the Crampon Translation (French))

"… formless void and darkness covered the face of the deep" (NRSV)

"… waste, and darkness was upon the face of the deep" (Dillman Translation)

"… waste and void, and darkness is on the face of the deep" (Robert Young Translation)

"… a desolate waste, with darkness covering the abyss" (S&G)

"… empty and had no form. Darkness covered the ocean" (EB)

"… desolate and empty with darkness on the face of the raging waters" (R. B. Thieme Translation)

"… waste and void, and darkness was upon the face of the deep" (Leupold Translation)

"… a formless waste, with darkness over the seas" (Speiser Translation)

"… formless void, there was darkness over the deep" (NJB)

"… without form or life, and darkness was upon the face of the deep" (Cassuto Translation)

"... formless and empty, darkness was over the surface of the deep" (NIV)

"... without form became waste, and void; and darkness was upon the face of the deep abyss." (ERRB)

"... unsightly and unfurnished, and darkness was over the deep" (Septuagint, Bagster Edition)

"... formless and void, and darkness was over the surface of the deep" (NASB)

"... waste and wild, and darkness was on the face of the roaring deep" (TEB)

"... formless and empty, and darkness was on the face of the deep." (SCRIP)

The words "without form, and void" are the Hebrew words *TOHU WAW BOHUW*, and "darkness" is *CHOSHEK*. Whatever these words mean, we can certainly tell that earth wasn't a very nice place. *Strong's Dictionary of the Hebrew Language*[18] defines them as:

TOHUW (H8414) "to'-hoo; from an unused root mean. to lie waste; a desolation (of surface), i.e. desert; fig. a worthless thing; adv. in vain:—confusion, empty place, without form, nothing, (thing of) nought, vain, vanity, waste, wilderness."

BOHUW (H922) "bo-hoo; from an unused root (mean. to be empty); a vacuity, i.e. (superficially) an undistinguishable ruin:—emptiness, void."

144

CHOSHEK (H2822) "kho-shek';... lit. darkness; fig. misery, destruction, death, ignorance, sorrow, wickedness: — dark (-ness), night, obscurity."

I've listed enough different translations for you to get a feeling for what these words mean. This is especially so for *CHOSHEK*, which is almost always translated "darkness." Only the Crampon translation has something different. I assume he believed a thick cloud-cover caused the earth to be dark. I won't argue whether or not he is correct. In fact, I'm not going to argue over any of these words except for one aspect of their meanings. The one factor we need to consider is whether these words should be translated in a literal sense or translated in a figurative sense? IT IS VERY IMPORTANT THAT WE BE ABLE TO TELL WHEN A WORD IS USED LITERALLY AND WHEN IT IS USED FIGURATIVELY. Our interpretations cannot be informationally correct if we don't know the difference between literal and figurative words. (Don't confuse figurative with untrue. If I say that something in the Bible is figurative, it does not mean that I am saying it is untrue. When Jesus compared Christians to wheat and unbelievers to tares, He was expressing truth in a figurative fashion. Christians aren't literal stalks of wheat with kernels of grain growing out of their heads. We do produce a spiritual fruit, however. While Jesus spoke truth, we must recognize that He wasn't being literal. If He was being literal, then there are no Christians in the world. No one has grain sprouting from his head. The Bible often uses figurative language, figures of speech, to explain truth.) Now look back at Strong's definitions of *TOHUW* and *CHOSHEK*. He lists both the literal and the figurative meanings of these words. He lists only a literal meaning for *BOHUW* because *BOHUW* is used only in its literal sense in the Bible. I'm going to list all the verses in the Bible that use these words because I want you to understand how they are used both literally and figuratively.

CHOSHEK

In its literal meaning, *CHOSHEK* designates physical darkness or the absence of light. If used figuratively, it can mean misery, destruction, death, ignorance, sorrow, or wickedness. In some places it is used literally, while in other places it is used figuratively. Here is how *CHOSEK* (in bold print) is translated in the *King James Bible*.

Gen. 1:2 "And the earth was without form, and void; and **darkness** *was* upon the face of the deep. And the Spirit of God moved upon the face of the waters."

Gen. 1:4-5 "And God saw the light, that *it was* good: and God divided the light from the **darkness**. *{5}* And God called the light Day, and the **darkness** he called Night. And the evening and the morning were the first day."

Gen. 1:18 "And to rule over the day and over the night, and to divide the light from the **darkness**: and God saw that *it was* good."

Exo. 10:21-22 "And the LORD said unto Moses, Stretch out thine hand toward heaven, that there may be **darkness** over the land of Egypt, even **darkness** *which* may be felt. *{22}* And Moses stretched forth his hand toward heaven; and there was a thick **darkness** in all the land of Egypt three days:"

Exo. 14:20 "And it came between the camp of the Egyptians and the camp of Israel; and it was a cloud and **darkness** *to* *them*, but it gave light by night *to these*: so that the one came not near the other all the night."

Deu. 4:11"And ye came near and stood under the mountain; and the mountain burned with fire unto the midst of heaven, with **darkness**, clouds, and thick darkness."

Deu. 5:23 "And it came to pass, when ye heard the voice out of the midst of the **darkness**, (for the mountain did burn with fire,) that ye came near unto me, *even* all the heads of your tribes, and your elders;"

Josh. 2:5 "And it came to pass *about the time* of shutting of the gate, when it was **dark**, that the men went out: whither the men went I wot not: pursue after them quickly; for ye shall overtake them."

1 Sam. 2:9 "He will keep the feet of his saints, and the wicked shall be silent in **darkness**; for by strength shall no man prevail."

2 Sam. 22:12 "And he made **darkness** pavilions round about him, dark waters, *and* thick clouds of the skies."

2 Sam. 22:29 "For thou *art* my lamp, O LORD: and the LORD will lighten my **darkness**."

Job 3:4-5 "Let that day be **darkness**; let not God regard it from above, neither let the light shine upon it. {5} Let **darkness** and the shadow of death stain it; let a cloud dwell upon it; let the blackness of the day terrify it."

Job 5:14 "They meet with **darkness** in the daytime, and grope in the noonday as in the night."

Job 10:21 "Before I go *whence* I shall not return, *even* to the land of **darkness** and the shadow of death;"

Job 12:22 "He discovereth deep things out of **darkness**, and bringeth out to light the shadow of death."

Job 12:25 "They grope in the **dark** without light, and he maketh them to stagger like *a* drunken *man*."

Job 15:22-23 "He believeth not that he shall return out of **darkness**, and he is waited for of the sword. *{23}* He wandereth abroad for bread, *saying*, Where *is it*? he knoweth that the day of **darkness** is ready at his hand."

Job 15:30 "He shall not depart out of **darkness**; the flame shall dry up his branches, and by the breath of his mouth shall he go away."

Job 17:12-13 "They change the night into day: the light *is* short because of **darkness**. *{13}* If I wait, the grave *is* mine house: I have made my bed in the **darkness**."

Job 18:18 "He shall be driven from light into **darkness**, and chased out of the world."

Job 19:8 "He hath fenced up my way that I cannot pass, and he hath set **darkness** in my paths."

Job 20:26 "All **darkness** *shall be* hid in his secret places: a fire not blown shall consume him; it shall go ill with him that is left in his tabernacle."

Job 22:11 "Or **darkness**, *that* thou canst not see; and abundance of waters cover thee."

Job 23:17 "Because I was not cut off before the **darkness**, *neither* hath he covered the darkness from my face."

Job 24:16 "In the **dark** they dig through houses, *which* they had marked for themselves in the daytime: they know not the light."

Job 26:10 "He hath compassed the waters with bounds, until the day and **night** come to an end."

Job 28:3 "He setteth an end to **darkness**, and searcheth out all perfection: the stones of **darkness**, and the shadow of death."

Job 29:3 "When his candle shined upon my head, *and when* by his light I walked *through* **darkness**;"

Job 34:22 *"There is* no **darkness**, nor shadow of death, where the workers of iniquity may hide themselves."

Job 37:19 "Teach us what we shall say unto him; *for* we cannot order *our speech* by reason of **darkness**."

Job 38:19 "Where is the way *where* light dwelleth? and *as for* **darkness**, where *is* the place thereof,"

Psa. 18:11 "He made **darkness** his secret place; his pavilion round about him *were* dark waters *and* thick clouds of the skies."

Psa. 18:28 "For thou wilt light my candle: the LORD my God will enlighten my **darkness**."

Psa. 35:6 "Let their way be **dark** and slippery: and let the angel of the LORD persecute them."

Psa. 88:12 "Shall thy wonders be known in the **dark**? and thy righteousness in the land of forgetfulness?"

Psa. 104:20 "Thou makest **darkness**, and it is night: wherein all the beasts of the forest do creep *forth.*"

Psa. 105:28 "He sent **darkness**, and made it dark; and they rebelled not against his word."

Psa. 107:10 "Such as sit in **darkness** and in the shadow of death, *being* bound in affliction and iron;"

Psa. 107:14 "He brought them out of **darkness** and the shadow of death, and brake their bands in sunder."

Psa. 112:4 "Unto the upright there ariseth light in the **darkness**: *he is* gracious, and full of compassion, and righteous."

Psa. 139:11-12 "If I say, Surely the **darkness** shall cover me; even the night shall be light about me. *{12}* Yea, the **darkness** hideth not from thee; but the night shineth as the day: the darkness and the light *are* both alike to *thee.*"

Prov. 2:13 "Who leave the paths of uprightness, to walk in the ways of **darkness**;"

Prov. 20:20 "Whoso curseth his father or his mother, his lamp shall be put out in obscure **darkness**."

Eccl. 2:13-14 "Then I saw that wisdom excelleth folly, as far as light excelleth **darkness**. *{14}* The wise man's eyes are in his head; but the fool walketh in **darkness**: and I myself perceived also that one event happeneth to them all."

Eccl. 5:17 "All his days also he eateth in **darkness**, and he hath much sorrow and wrath with his sickness."

Eccl. 6:4 "For he cometh in with vanity, and departeth in **darkness**, and his name shall be covered with **darkness**."

Eccl. 11:8 "But if a man live many years, *and* rejoice in them all; yet let him remember the days of **darkness**; for they shall be many. All that cometh *is* vanity."

Isa. 5:20 "Woe unto them that call evil good, and good evil; that put **darkness** for light, and light for **darkness**; that put bitter for sweet, and sweet for bitter!"

Isa. 5:30 "And in that day they shall roar against them like the roaring of the sea: and if *one* look unto the land, behold **darkness** *and* sorrow, and the light is darkened in the heavens thereof."

Isa. 9:2 "The people that walked in **darkness** have seen a great light: they that dwell in the land of the shadow of death, upon them hath the light shined."

Isa. 29:18 "And in that day shall the deaf hear the words of the book, and the eyes of the blind shall see out of obscurity, and out of **darkness**."

Isa. 42:7 "To open the blind eyes, to bring out the prisoners from the prison, *and* them that sit in **darkness** out of the prison house."

Isa. 45:3 "And I will give thee the treasures of **darkness**, and hidden riches of secret places, that thou mayest know that I, the LORD, which call *thee* by thy name, *am* the God of Israel."

Isa. 45:7 "I form the light, and create **darkness**: I make peace, and create evil: I the LORD do all these *things*."

Isa. 45:19 "I have not spoken in secret, in a **dark** place of the earth: I said not unto the seed of Jacob, Seek ye me in vain: I the LORD speak righteousness, I declare things that are right."

Isa. 47:5 "Sit thou silent, and get thee into **darkness**, O daughter of the Chaldeans: for thou shalt no more be called, The lady of kingdoms."

Isa. 49:9 "That thou mayest say to the prisoners, Go forth; to them that *are* in **darkness**, Show yourselves. They shall feed in the ways, and their pastures *shall be* in all high places."

Isa. 58:10 "And *if* thou draw out thy soul to the hungry, and satisfy the afflicted soul; then shall thy light rise in obscurity, and thy **darkness** *be* as the noon day:"

Isa. 59:9 "Therefore is judgment far from us, neither doth justice overtake us: we wait for light, but behold obscurity; for brightness, *but* we walk in **darkness**."

Isa. 60:2 "For, behold, the **darkness** shall cover the earth, and gross darkness the people: but the LORD shall arise upon thee, and his glory shall be seen upon thee."

Lam. 3:2 "He hath led me, and brought *me into* **darkness**, but not *into* light."

Eze. 8:12 "Then said he unto me, Son of man, hast thou seen what the ancients of the house of Israel do in the **dark**, every man in the chambers of his imagery? for they say, The LORD seeth us not; the LORD hath forsaken the earth."

Eze. 32:8 "All the bright lights of heaven will I make dark over thee, and set **darkness** upon thy land, saith the Lord GOD."

Joel 2:2 "A day of **darkness** and of gloominess, a day of clouds and of thick darkness, as the morning spread upon the mountains: a great people and a strong; there hath not been ever the like, neither shall be any more after it, *even* to the years of many generations."

Joel 2:31 "The sun shall be turned into **darkness**, and the moon into blood, before the great and the terrible day of the LORD come."

Amos 5:18 "Woe unto you that desire the day of the LORD! to what end *is* it for you? the day of the LORD *is* **darkness**, and not light."

Amos 5:20 "*Shall* not the day of the LORD *be* **darkness**, and not light? even very dark, and no brightness in it?"

Micah 7:8 "Rejoice not against me, O mine enemy: when I fall, I shall arise; when I sit in **darkness**, the LORD *shall be* a light unto me."

Nahum 1:8 "But with an overrunning flood he will make an utter end of the place thereof, and **darkness** shall pursue his enemies."

Zep. 1:15 "That day *is* a day of wrath, a day of trouble and distress, a day of wasteness and desolation, a day of **darkness** and gloominess, a day of clouds and thick darkness,"

It's fairly evident when *CHOSHEK* is literal and when it's figurative. The context of Genesis One is a physical description of the earth. When *CHOSHEK* describes a physical condition, it is used in its literal sense. Genesis 1:2 is not a figurative image. The earth was literally, physically dark and without light before day one.

King James Version Translations of *TOHUW* and *BOHUW*

TOHUW is used twenty times. *BOHUW* is used three times. *TOHUW* literally describes physical condition, but figuratively, it describes lack of value or lack of purpose. *BOHUW* is always literal. The bold words are translations of **TOHUW** and the underlined words are translations of BOHUW. The three references in bold and underlined are the verses with both.

Gen. 1:2 "And the earth was **without form**, and void; and darkness *was* upon the face of the deep. And the Spirit of God moved upon the face of the waters."

Deu. 32:10 "He found him in a desert land, and in the **waste** howling wilderness; he led him about, he instructed him, he kept him as the apple of his eye."

1 Sam. 12:21 "And turn ye not aside: for *then should ye* go after **vain** *things*, which cannot profit nor deliver; for they *are* **vain**."

Job 6:18 "The paths of their way are turned aside; they go to **nothing**, and perish."

Job 12:24 "He taketh away the heart of the chiefs if the people of the earth, and causeth them to wander in a **wilderness** *where there is* no way."

Job 26:7 "He stretcheth out the north over the **empty place**, *and* hangeth the earth upon nothing."

Psa. 107:40 "He poureth contempt upon princes, and causeth them to wander in the **wilderness**, *where there is* no way."

Isa. 24:10 "The city of **confusion** is broken down: every house is shut up, that no man may come in."

Isa. 29:21 "That make a man an offender for a word, and lay a snare for him that reproveth in the gate, and turn aside the just for a **thing of nought**."

Isa. 34:11 "But the cormorant and the bittern shall possess it; the owl also and the raven shall dwell in it: and he shall stretch out upon it the line of **confusion**, and the stones of emptiness."

Isa. 40:17 "All nations before him *are* as nothing; and they are counted to him less than nothing, and **vanity**."

Isa. 40:23 "That bringeth the princes to nothing; he maketh the judges of the earth as **vanity**."

Isa. 41:29 "Behold, they *are* all vanity; their works *are* nothing: their molten images *are* wind and **confusion**."

Isa. 44:9 "They that make a graven image *are* all of them **vanity**; and their delectable things shall not profit; and they *are* their own witnesses; they see not, nor know; that they may be ashamed."

Isa. 45:18-19 "For thus saith the LORD that created the heavens; God himself that formed the earth and made it; he hath established it, he created it not **in vain**, he formed it to be inhabited: I *am* the LORD; and *there is* none else. *{19}* I have not spoken in secret, in a dark place of the earth: I said not unto the seed of Jacob, Seek ye me **in vain**: I the LORD speak righteousness, I declare things that are right."

Isa. 49:4 "Then I said, I have laboured in vain, I have spent my strength **for nought**, and in vain: *yet* surely my judgment is with the LORD, and my work with my God."

Isa. 59:4 "None calleth for justice, nor *any* pleadeth for truth: they trust **in vanity**, and speak lies; they conceive mischief, and bring forth iniquity."

Jer. 4:23 "I beheld the earth, and, lo, *it was* **without form**, and void; and the heavens, and they *had* no light."

BOHUW means void or empty. The Bible always uses the literal meaning of *BOHUW*. *TOHUW*, on the other hand, can be either literal or figurative. In its literal meaning *TOHUW* describes a desolation, a barren wasteland, or a bleak wilderness. *TOHUW* figuratively describes vanity, purposelessness, nonproductiveness, worthlessness, or uselessness. So what does *TOHUW* mean in Genesis 1:2? Genesis 1:2 is giving us the physical condition of the earth. We are seeing earth as it literally was. It was a desolation, a wasteland. Some describe it as a chaos, if such a thing can exist, and it was uninhabited. There is nothing figurative in Genesis One. Genesis One is describing the physical creation. There is nothing hinting that Genesis 1:2 should be interpreted in a figurative sense. To translate *TOHUW* figuratively here would mean that God had no use or purpose or reason for creating earth. It might be grammatically correct to translate it figuratively, but such a translation would mean that God had no purpose for creating earth. It would imply that earth's existence had no meaning, value, or worth. It would indicate that God had no use for the earth, or that creating it was something He did in vain. This interpretation makes the Bible disagree with itself.

Jer. 51:15 "It is he who made the earth by his power, who established the works by his wisdom, and by his understanding stretched out the heavens." (NRSV)

Clearly the Bible indicates that God had carefully planned His creation; it was no useless, pointless, thoughtless act. Therefore, *TOHUW* must be translated in its literal sense. *TOHUW* describes the physical condition of the earth at the beginning of Genesis 1:2. Not only was it *TOHUW*, a desolation, it was also *BOHUW*, empty. Most likely this is referring to being empty of life. Genesis 1:2 describes earth as dark, desolate, and dead, but Isaiah 45 seems to indicate He didn't create it that way. Genesis 1:3-31 describe what God did to that dark, desolate, and dead planet. It is important to understand that Genesis 1:3-31 describe these events from an earthly perspective. When we read His account of His creation, we are not looking down from heaven. We are looking at things as if we were standing on the earth watching God work. I now ask the question, "Do these words describe the earth when God first created it, or did the earth become dark, desolate, and dead later?" What we need to decide is whether there is a break, a disconnect, between the creation of the earth in Genesis 1:1 and the dark, desolate, and dead earth in Genesis 1:2. The only way we can know how to translate Genesis 1:1-2 is to know what the rest of the Bible reveals about the creation.

Chapter Seven:

The Restoration Theory

—◈—

I believe God originally created the universe out of nothing at some distant time in the past. Matter and energy did not coexist with God. At that instant of "timelessness," God created space, time, energy, and matter. I don't think He created the universe with galaxies, stars, and planets intact; I think He created them later. Now, some believe that Genesis 1:1 indicates they were created intact and together. However, "created" doesn't necessarily imply an instantaneous point of time. God has a funny way of describing events that happen over long periods of time as if they happen together. (I guess it's an occupational hazard of being omniscient, omnipresent, and eternal.) Grammatically, Genesis 1:1 could describe the instantaneous creation of an intact earth, but there is a passage in the Book of Job that seems to indicate otherwise.

Job 38:4-6 "Where were you when I laid the foundation of the earth? Tell me, if you have understanding. {5} Who determined its measurements—surely you know! Or who stretched the line upon it? {6} On what were its bases sunk, or who laid its cornerstone..." (NRSV)

Here we have God telling Job something about the creation of the earth. God is revealing information to Job that He didn't reveal to Moses. In fact, some scholars believe that Job lived before Moses and the Book of Job is older than the Pentateuch. That's an issue for others to argue. What I want to focus on is what was revealed. God used some imagery, some figurative language, to describe the earth's creation. God mentions a foundation and the laying of that foundation. He talks about laying a corner stone. He talks about stretching the line on it and measuring it. He asks Job about the bases that were sunk. All this seems to imply that the earth was constructed, as a building would be, over some period of time. Laying a foundation, laying out its measures, stretching a line upon it, laying a cornerstone, and sinking its bases would have suggested a process to Job. Job understood what constructing a building involves. It takes raw materials and it takes time. Why would God suggest a process if He had created the earth instantaneously? Now don't go goofy on me and start using this to defend a Flat Earth. This doesn't mean we're supposed to believe the earth is sitting on a big pile of rocks. These images are figurative. Earth has no foundation; it's floating in space. It has no corner stone; it's a sphere. We know this from scientific observations. (True science filters out false theology.) If we took these words to be literal descriptions of the earth, then we'd have to believe the earth was made of giant stones built on a foundation. Since these words are figurative, we don't have to defend a doctrine that says the earth is lying on a foundation of stones. Nevertheless, God is trying to reveal something to Job by using imagery. What was He trying to reveal? **Throughout the Bible, when God uses imagery, figurative language, similes, metaphors, parables, hyperbole, or whatever, it is designed to reveal truth.** God doesn't use imagery to convey nothing. In fact, the use of imagery, such as Jesus teaching in parables, teaches far more than stating raw facts alone. No, we really aren't sheep, but like sheep we are helpless, hopeless, stupid, prone to wander, and from the perspective of God's Holy character, we stink! While the image of us being sheep may hurt our egos, it is not misleading. It reveals the truth about us to us. The same must be true for the imagery God uses here in Job. God is using imagery to teach us something. If the earth wasn't created by means of a process over time, then this imagery is misleading. God should have asked Job, "Where were you when I created the earth...," and left it at that. He didn't have to suggest a process if there was no process. In fact, He shouldn't have suggested a process if there was no process. There are two reasons why we should worry about God suggesting things that aren't true. The first is that it means God can add false information and concepts to the Bible. That's not good. The second is that it means we can't trust the Bible's teachings about Jesus Christ. If God's imagery about **laying the foundations** of the earth is misleading, then God's imagery about Jesus, the Good Shepherd **laying down His life** for His sheep, can also be misleading. We can't trust that imagery either. That's worse! As I look at this passage in Job, I can't help but interpret it to mean that the creation of the earth was a time-consuming process and not instantaneous.

Let me now guess what some will say. They will say that this contradicts the doctrine of *ex nihilo* creation. It doesn't. As I said before, creation out of nothing does not imply instantaneous creation. If God created all the raw materials out of nothing and then fashioned them into a universe over time, it would still be *ex nihilo*. I think this is why the Bible uses three separate words to describe God's creative actions: *BARA* (create), *ASAH* (make), and *YATSAR* (form). "Okay, Steve," they'll say, "but why can't that time-consuming process be the creative acts of God during the six days of creation? That was a process over time. What makes you think God wasn't telling Job about those six days?" The answer is that the two accounts, Job 38:4-6 and Genesis 1:3-31, describe different aspects of earth's creation. They don't describe the same events. In Job, God describes the earth itself. He talks about its foundation, its bases, and its corner stone. If I had to put this in modern terms, I would suggest that God was talking about the earth's core, its mantle, its crust, and its strata. The verses in Job describe the creation of the earth from the surface down. Genesis 1:3-31 describe the atmosphere, the oceans, the clouds, the trees, the birds, and all the rest. These verses describe what God did from the surface up. Yes, that was a process over time too, but it was a different process at a different time.

I believe Genesis 1:1 describes the original creation of the heavens and earth. I don't know when the earth was created in relation to the rest of the universe, but I know it wasn't created a desolate waste! It wasn't until after sin entered the universe, through Lucifer, that something terrible happened to the earth. Lucifer began corrupting all of creation, both heaven and earth. All of creation "groans" because of his sin. I think he began defiling the earth and the life on it. I believe God intervened and judged the earth, leaving it dead and dark for an unspecified (but probably very short) period of time. I believe Genesis 1:3-31 describe the restoration. This was not the creation of a New Heavens and a New Earth. The New Heavens and the New Earth are future creations. This was a restoration of the same heavens and the same earth that had existed before. I don't think everything was fully restored to what it had been, but it was restored enough for man's needs. According to the Restoration Theory, Pre-Adamic life on earth was destroyed by God's judgment. Following its *TOHUW WAW BOHUW* period, the earth was restored by God during six literal days. According to the Restoration Theory, all living things today are descendants of the living things He restored to the earth during those six days; nothing evolved.

Before I continue, let me say one thing. If you don't like my opinion, that's fine. But don't criticize it on the basis that it has no grammatical support or that scholars don't accept it. If anyone tells you this, they are wrong. I absolutely, positively guarantee that many scholars down through the ages have defended the Restoration Theory. The idea of a Pre-Adamic global catastrophe is thousands of years old. In spite of that, many Christian Creationists claim that Thomas Chalmers invented the Gap Theory in 1814 to make the Bible fit Darwin's Theory of Evolution. It disturbs me that Christians who know the truth still say things they know aren't true. Charles Darwin was five years old when Chalmers began defending the Gap Theory. If that was Chalmers' motive, then he was a prophet. Why do people still say and believe things they know aren't true? It brings up the question of motives. Today, both Young-Earth and Day-Age creationists reject the Gap Theory. Their reasons are different, but their motives are identical. Their rejection of the Restoration Theory brings to light a much darker motive for their beliefs.

Young-Earth creationists say the six days were literal days. In their minds, if the days were literal, then the universe cannot be old. This is why they reject the Gap Theory. The Gap Theory agrees that the days were literal days, but Young-Earthers still reject it because it goes contrary to their most cherished belief, the belief that the universe is young. Young-Earth creationists are not as concerned with WHAT the days were, as they are with WHEN the days were. Young-Earth creationists care more about the earth being young than they do about the days being literal. They spend virtually all their time, talent, and money on trying to prove the universe is only a few thousand years old. Too many doctorates have been earned, too much money has been spent, and too many reputations are on the line to permit them to accept a theory that allows the universe to be old. Allowing that would mean they have put their own prestige above what God wants revealed in His Word. Now, I believe God revealed what He wants us to know. What does He reveal in the very first chapter of the very first book of the Bible? God reveals that the six days were literal days, but He doesn't say a thing about when those six days were. God could have told

Moses the exact time it happened, but He didn't. God did not tell Moses whether it was a few thousand years before or several billion years before. God's emphasis is on WHAT the days were, not on WHEN the days were. Young-Earth creationists reverse the emphasis. The age of the universe is more important. **If the days being literal was more important, then they wouldn't attack the Gap Theory.**

Day-Age creationists have the same motive for rejecting the Restoration Theory, but their reason is different. In their case, they can't let the days be anything but ages. By their thinking, if the universe is old, then the days must be ages. So, like the Young-Earthers, they reject the Gap Theory. The Gap Theory allows for the universe to be billions of years old just like the Day-Age Theory. Day-Age creationists still reject it because it says the days were literal. The Gap Theory attacks their most cherished belief, the belief that the days are ages. **If the age of the universe was more important, then they wouldn't attack the Gap Theory.** Instead, they have put too much effort, written too many books, and given too many lectures emphasizing that the days are ages. If they admit that the days were literal, then they would have to admit that their efforts were misguided. This is why they must insist, like the Young-Earth creationists, that *HAYAH* cannot be translated "became." They have the same motive. Personal prestige, reputation, and ego trump God's revelation. Ironically, both Day-Age creationists and Young-Earth creationists commit the same error they accuse evolutionists of committing. Namely, they have painted themselves into philosophical corners, and can't escape without admitting they were wrong. No one wants to look foolish, so no one budges from his corner. Their own clever theories become more important than what the evidence shows.

Because of this, both camps say that *HAYAH* should not or can not be translated "became." What does the Bible say? *HAYAH* is translated "became," "become," "came," "came to pass," and "come to pass" about 800 times in the Bible. In the *King James Version*, *HAYAH* is translated this way approximately 90 times in Genesis alone.

KJV Translations of *HAYAH* (in bold print) in Genesis as Some Form of "Became"

Gen. 2:7 "And the LORD God formed man of the dust of the ground, and breathed into his nostrils the breath of life; and man **became** a living soul."

Gen. 2:10 "And a river went out of Eden to water the garden; and from thence it was parted, and **became** into four heads."

Gen. 3:22 "And the LORD God said, Behold, the man is **become** as one of us, to know good and evil: and now, lest he put forth his hand, and take also of the tree of life, and eat, and live for ever:"

Gen. 4:3 "And in process of time it **came to pass**, that Cain brought of the fruit of the ground an offering unto the LORD."

Gen. 4:8 "And Cain talked with Abel his brother: and it **came to pass**, when they were in the field, that Cain rose up against Abel his brother, and slew him."

Gen. 4:14 "Behold, thou hast driven me out this day from the face of the earth; and from thy face shall I be hid; and I shall be a fugitive and a vagabond in the earth; and it shall **come to pass**, that every one that findeth me shall slay me."

Gen. 6:1 "And it **came to pass**, when men began to multiply on the face of the earth, and daughters were born unto them,"

Gen. 7:10 "And it **came to pass** after seven days, that the waters of the flood were upon the earth."

Gen. 8:6 "And it **came to pass** at the end of forty days, that Noah opened the window of the ark which he had made:"

Gen. 8:13 "And it **came to pass** in the six hundredth and first year, in the first month, the first day of the month, the waters were dried up from off the earth: and Noah removed the covering of the ark, and looked, and, behold, the face of the ground was dry."

Gen. 9:14-15 "And it shall **come to pass**, when I bring a cloud over the earth, that the bow shall be seen in the cloud: {15} And I will remember my covenant, which is between me and you and every living creature of all flesh; and the waters shall no more **become** a flood to destroy all flesh."

Gen. 11:2 "And it **came to pass**, as they journeyed from the east, that they found a plain in the land of Shinar; and they dwelt there."

Gen. 12:11 "And it **came to pass**, when he was come near to enter into Egypt, that he said unto Sarai his wife, Behold now, I know that thou art a fair woman to look upon:"

Gen. 12:14 "And it **came to pass**, that, when Abram was come into Egypt, the Egyptians beheld the woman that she was very fair."

Gen. 14:1 "And it **came to pass** in the days of Amraphel king of Shinar, Arioch king of Ellasar, Chedorlaomer king of Elam, and Tidal king of nations;"

Gen. 15:1 "After these things the word of the LORD **came** unto Abram in a vision, saying, Fear not, Abram: I am thy shield, and thy exceeding great reward."

Gen. 15:17 "And it **came to pass**, that, when the sun went down, and it was dark, behold a smoking furnace, and a burning lamp that passed between those pieces."

Gen. 18:18 "Seeing that Abraham shall surely **become** a great and mighty nation, and all the nations of the earth shall be blessed in him?"

Gen. 19:17 "And it **came to pass**, when they had brought them forth abroad, that he said, Escape for thy life; look not behind thee, neither stay thou in all the plain; escape to the mountain, lest thou be consumed."

Gen. 19:26 "But his wife looked back from behind him, and she **became** a pillar of salt."

Gen. 19:29 "And it **came to pass**, when God destroyed the cities of the plain, that God remembered Abraham, and sent Lot out of the midst of the overthrow, when he overthrew the cities in the which Lot dwelt."

Gen. 19:34 "And it **came to pass** on the morrow, that the firstborn said unto the younger, Behold, I lay yesternight with my father: let us make him drink wine this night also; and go thou in, and lie with him, that we may preserve seed of our father."

Gen. 20:12-13 "And yet indeed she is my sister; she is the daughter of my father, but not the daughter of my mother; and she **became** my wife. {13} And it **came to pass**, when God caused me to wander from my father's house, that I said unto her, This is thy kindness which thou shalt show unto me; at every place whither we shall come, say of me, He is my brother."

Gen. 21:20 "And God was with the lad; and he grew, and dwelt in the wilderness, and **became** an archer."

Gen. 21:22 "And it **came to pass** at that time, that Abimelech and Phichol the chief captain of his host spake unto Abraham, saying, God is with thee in all that thou doest:"

Gen. 22:1 "And it **came to pass** after these things, that God did tempt Abraham, and said unto him, Abraham: and he said, Behold, here I am."

Gen. 22:20 "And it **came to pass** after these things, that it was told Abraham, saying, Behold, Milcah, she hath also born children unto thy brother Nahor;

Gen. 24:14-15 "And let it **come to pass**, that the damsel to whom I shall say, Let down thy pitcher, I pray thee, that I may drink; and she shall say, Drink, and I will give thy camels drink also: let the same be she that thou hast appointed for thy servant Isaac; and thereby shall I know that thou hast showed kindness unto my master. {15} And it **came to pass**, before he had done speaking, that, behold, Rebekah came out, who was born to Bethuel, son of Milcah, the wife of Nahor, Abraham's brother, with her pitcher upon her shoulder."

Gen. 24:22 "And it **came to pass**, as the camels had done drinking, that the man took a golden earring of half a shekel weight, and two bracelets for her hands of ten shekels weight of gold;"

Gen. 24:30 "And it **came to pass**, when he saw the earring and bracelets upon his sister's hands, and when he heard the words of Rebekah his sister, saying, Thus spake the man unto me; that he came unto the man; and, behold, he stood by the camels at the well."

Gen. 24:43 "Behold, I stand by the well of water; and it shall **come to pass**, that when the virgin cometh forth to draw water, and I say to her, Give me, I pray thee, a little water of thy pitcher to drink;"

Gen. 24:52 "And it **came to pass**, that, when Abraham's servant heard their words, he worshipped the LORD, bowing himself to the earth."

Gen. 24:67 "And Isaac brought her into his mother Sarah's tent, and took Rebekah, and she **became** his wife; and he loved her: and Isaac was comforted after his mother's death."

Gen. 25:11 "And it **came to pass** after the death of Abraham, that God blessed his son Isaac; and Isaac dwelt by the well Lahairoi."

Gen. 26:8 "And it **came to pass**, when he had been there a long time, that Abimelech king of the Philistines looked out at a window, and saw, and, behold, Isaac was sporting with Rebekah his wife."

Gen. 26:32 "And it **came to pass** the same day, that Isaac's servants came, and told him concerning the well which they had digged, and said unto him, We have found water."

Gen. 27:1 "And it **came to pass**, that when Isaac was old, and his eyes were dim, so that he could not see, he called

Esau his eldest son, and said unto him, My son: and he said unto him, Behold, here am I."

Gen. 27:30 "And it **came to pass**, as soon as Isaac had made an end of blessing Jacob, and Jacob was yet scarce gone out from the presence of Isaac his father, that Esau his brother came in from his hunting."

Gen. 27:40 "And by thy sword shalt thou live, and shalt serve thy brother; and it shall **come to pass** when thou shalt have the dominion, that thou shalt break his yoke from off thy neck."

Gen. 29:10 "And it **came to pass**, when Jacob saw Rachel the daughter of Laban his mother's brother, and the sheep of Laban his mother's brother, that Jacob went near, and rolled the stone from the well's mouth, and watered the flock of Laban his mother's brother."

Gen. 29:13 "And it **came to pass**, when Laban heard the tidings of Jacob his sister's son, that he ran to meet him, and embraced him, and kissed him, and brought him to his house. And he told Laban all these things."

Gen. 29:23 "And it **came to pass** in the evening, that he took Leah his daughter, and brought her to him; and he went in unto her."

Gen. 29:25 "And it **came to pass**, that in the morning, behold, it was Leah: and he said to Laban, What is this thou hast done unto me? did not I serve with thee for Rachel? wherefore then hast thou beguiled me?"

Gen. 30:25 "And it **came to pass**, when Rachel had born Joseph, that Jacob said unto Laban, Send me away, that I may go unto mine own place, and to my country."

Gen. 30:41 "And it **came to pass,** whensoever the stronger cattle did conceive, that Jacob laid the rods before the eyes of the cattle in the gutters, that they might conceive among the rods."

Gen. 31:10 "And it **came to pass** at the time that the cattle conceived, that I lifted up mine eyes, and saw in a dream, and, behold, the rams which leaped upon the cattle were ringstreaked, speckled, and grisled."

Gen. 32:10 "I am not worthy of the least of all the mercies, and of all the truth, which thou hast showed unto thy servant; for with my staff I passed over this Jordan; and now I am **become** two bands."

Gen. 34:16 "Then will we give our daughters unto you, and we will take your daughters to us, and we will dwell with you, and we will **become** one people."

Gen. 34:25 "And it **came to pass** on the third day, when they were sore, that two of the sons of Jacob, Simeon and Levi, Dinah's brethren, took each man his sword, and came upon the city boldly, and slew all the males."

Gen. 35:17-18 "And it **came to pass**, when she was in hard labour, that the midwife said unto her, Fear not; thou shalt have this son also. *{18}* And it **came to pass**, as her soul was in departing, (for she died) that she called his name Benoni: but his father called him Benjamin."

Gen. 35:22 "And it **came to pass**, when Israel dwelt in that land, that Reuben went and lay with Bilhah his father's concubine: and Israel heard it. Now the sons of Jacob were twelve:"

Gen. 37:20 "Come now therefore, and let us slay him, and cast him into some pit, and we will say, Some evil beast hath devoured him: and we shall see what will **become** of his dreams."

Gen. 37:23 "And it **came to pass**, when Joseph was come unto his brethren, that they stripped Joseph out of his coat, his coat of many colours that was on him;"

Gen. 38:1 "And it **came to pass** at that time, that Judah went down from his brethren, and turned in to a certain Adullamite, whose name was Hirah."

Gen. 38:9 "And Onan knew that the seed should not be his; and it **came to pass**, when he went in unto his brother's wife, that he spilled it on the ground, lest that he should give seed to his brother."

Gen. 38:24 "And it **came to pass** about three months after, that it was told Judah, saying, Tamar thy daughter in law hath played the harlot; and also, behold, she is with child by whoredom. And Judah said, Bring her forth, and let her be burnt."

Gen. 38:27-29 "And it **came to pass** in the time of her travail, that, behold, twins were in her womb. {28} And it **came to pass**, when she travailed, that the one put out his hand: and the midwife took and bound upon his hand a scarlet thread, saying, This came out first. {29} And it **came to pass**, as he drew back his hand, that, behold, his brother came out: and she said, How hast thou broken forth? this breach be upon thee: therefore his name was called Pharez."

Gen. 39:5 "And it **came to pass** from the time that he had made him overseer in his house, and over all that he had, that the LORD blessed the Egyptian's house for Joseph's sake;

and the blessing of the LORD was upon all that he had in the house, and in the field."

Gen. 39:7 "And it **came to pass** after these things, that his master's wife cast her eyes upon Joseph; and she said, Lie with me."

Gen. 39:10-11 "And it **came to pass**, as she spake to Joseph day by day, that he hearkened not unto her, to lie by her, or to be with her. *{11}* And it **came to pass** about this time, that Joseph went into the house to do his business; and there was none of the men of the house there within."

Gen. 39:13 "And it **came to pass**, when she saw that he had left his garment in her hand, and was fled forth,"

Gen. 39:15 "And it **came to pass**, when he heard that I lifted up my voice and cried, that he left his garment with me, and fled, and got him out."

Gen. 39:18-19 "And it **came to pass**, as I lifted up my voice and cried, that he left his garment with me, and fled out. *{19}* And it **came to pass**, when his master heard the words of his wife, which she spake unto him, saying, After this manner did thy servant to me; that his wrath was kindled."

Gen. 40:1 "And it **came to pass** after these things, that the butler of the king of Egypt and his baker had offended their lord the king of Egypt."

Gen. 40:20 "And it **came to pass** the third day, which was Pharaoh's birthday, that he made a feast unto all his servants: and he lifted up the head of the chief butler and of the chief baker among his servants."

Gen. 41:1 "And it **came to pass** at the end of two full years, that Pharaoh dreamed: and, behold, he stood by the river."

Gen. 41:8 "And it **came to pass** in the morning that his spirit was troubled; and he sent and called for all the magicians of Egypt, and all the wise men thereof: and Pharaoh told them his dream; but there was none that could interpret them unto Pharaoh."

Gen. 41:13 "And it **came to pass**, as he interpreted to us, so it was; me he restored unto mine office, and him he hanged."

Gen. 42:35 "And it **came to pass** as they emptied their sacks, that, behold, every man's bundle of money was in his sack: and when both they and their father saw the bundles of money, they were afraid."

Gen. 43:2 "And it **came to pass**, when they had eaten up the corn which they had brought out of Egypt, their father said unto them, Go again, buy us a little food."

Gen. 43:21 "And it **came to pass**, when we came to the inn, that we opened our sacks, and, behold, every man's money was in the mouth of his sack, our money in full weight: and we have brought it again in our hand."

Gen. 44:24 "And it **came to pass** when we came up unto thy servant my father, we told him the words of my lord."

Gen. 44:31 "It shall **come to pass**, when he seeth that the lad is not with us, that he will die: and thy servants shall bring down the gray hairs of thy servant our father with sorrow to the grave."

Gen. 46:33 "And it shall **come to pass**, when Pharaoh shall call you, and shall say, What is your occupation?"

Gen. 47:20 "And Joseph bought all the land of Egypt for Pharaoh; for the Egyptians sold every man his field, because the famine prevailed over them: so the land **became** Pharaoh's."

Gen. 47:24 "And it shall **come to pass** in the increase, that ye shall give the fifth part unto Pharaoh, and four parts shall be your own, for seed of the field, and for your food, and for them of your households, and for food for your little ones."

Gen. 47:26 "And Joseph made it a law over the land of Egypt unto this day, that Pharaoh should have the fifth part; except the land of the priests only, which **became** not Pharaoh's."

Gen. 48:1 "And it **came to pass** after these things, that one told Joseph, Behold, thy father is sick: and he took with him his two sons, Manasseh and Ephraim."

Gen. 48:19 "And his father refused, and said, I know it, my son, I know it: he also **shall become** a people, and he also shall be great: but truly his younger brother shall be greater than he, and his seed **shall become** a multitude of nations."

Gen. 49:15 "And he saw that rest was good, and the land that it was pleasant; and bowed his shoulder to bear, and **became** a servant unto tribute."

Can *HAYAH* denote "becoming" rather than "being?" Absolutely! Are the Young-Earth and Day-Age statements about *HAYAH* facts? No, they are opinions, and very biased ones at that. Using *The Complete Word Study Old Testament King James Version*[19], I counted 315 *HAYAH*s in Genesis.

(See Chapter Twelve for more details about *HAYAH*.) Since I did this count manually and not by computer, I'm willing to admit I may have missed a few, but I don't think I missed many. Since *HAYAH* is translated as some form of "became" about one-third of the time, I fail to see how anyone can claim that *HAYAH* can't be translated as some form of "became." Personal opinion has a way of becoming proven fact given enough time and ignorance. This is why I emphasize that we must be careful when people make themselves appear scholarly. I have never objected to personal opinions as long as they are so labeled. However, when opinions are proclaimed as facts and defended by untrue statements, I get angry. I get even angrier if Christians do it. **No lie, no matter how eloquently expressed, will ever glorify Jesus Christ.** My warning to you is to be very careful what you read and whom you believe, especially if a non-Hebrew or Greek scholar tries to convince you that he is one. Once again, I openly admit I am not a scholar in Hebrew or Greek. However, the Restoration Theory is a theory that has been believed and defended by a number of excellent Hebrew and Greek scholars. For me, it is simply the theory that seems to fit all the facts best.

How My Thinking Evolved

Even before I became a creationist, I was aware of the Restoration Theory. I first read about it in *The New Scofield Reference Bible*.[20] Dr. C. I. Scofield (1843-1921), certainly a qualified biblical scholar, preferred the Restoration Theory. I was intrigued by the idea of a prior earth and a later restoration, but I remained unconvinced. At the time, I was still a theistic evolutionist. Later, I purchased a copy of the eye-opening book, *Dispensational Truth*,[21] by Clarence Larkin (1850-1924). He believed in the Restoration Theory as well. I was entertained by his presentation, especially by

175

his diagrams and drawings, but I wasn't convinced. I clung tightly to theistic evolution. It wasn't until after I was challenged with some Bible verses about man being created before woman that I began to suspect something was wrong with evolution. Once I finished my studies in science and discovered that evolution was a lie, I returned to the Bible to see what it truly said about our origin. I tried to use the same logical approach to determine truth in the Bible as I had used to determine truth from science. I wanted to let the Bible speak for itself. My motive was not to force the Bible to defend or defeat any particular creation theory. I wasn't even trying to force the Bible to defend creationism. That was the very thing I was attempting to discover. This was my goal: I wanted to see if the Bible could be blended with evolution. I knew that if the Bible taught evolution, then the Bible was wrong and Jesus was a hoax. I was so strongly convinced **from the scientific evidence** that evolution was a lie, that if the Bible did teach evolution, then I was ready to toss Christianity into the waste basket and start searching for my true Creator. After seeing for myself that evolution wasn't taught in the Bible, I opted for the Young-Earth Theory. I believed the universe, the earth, and life on the earth were originally created only a few thousand years ago during a six-day period.

I originally questioned my Day-Age, Theistic-Evolution belief because I was confronted with some passages of Scripture that couldn't be reconciled with evolution. In the same fashion, I questioned my Young-Earth belief when I was confronted with some passages of Scripture that couldn't be reconciled with the Young-Earth Theory. As I studied the Bible, I found some problem passages that didn't fit. The passage in Job 38 was one of them; I soon found others. Because of this, I began studying the different creation theories. I slowly began to understand how the Restoration Theory explained other problem biblical statements concerning the creation. It reconciled the Genesis account with the parallel creation accounts mentioned elsewhere in the Bible. Whether or not it made scientific sense wasn't an issue with me at the time. I was trying to discover what the Bible truly said. Once I realized that this was what God had revealed, I went back and looked at the science of the Restoration Theory. By that time I was firmly a creationist, so I certainly wasn't trying to find evidence to defend evolution. Some creationists proclaim that if you believe in the Restoration Theory, you are automatically an evolutionist. This is a wrong conclusion. I believe in the Restoration Theory but I totally reject evolution. There was no evolution going on in the time period between Genesis 1:1 and Genesis 1:3. There was no evolution going on during the time that God made the earth *TOHUW WAW BOHUW.* Genesis 1:2 says the earth was dead. There could be no evolution because there was no life. God tells us that the earth was a desolate waste, devoid of life and totally uninhabitable. It was dark and covered with water. Nothing could live. Evolution never happened. Life on the earth was originally created by God. Life on earth died sometime later. Life on earth was again created by God during the six twenty-four hour days described in the Book of Genesis. The Restoration Theory cannot be used to blend evolution with the Bible. Life did not evolve during the gap. Nothing could evolve because everything was dead. Even evolutionists agree that dead things don't evolve very well. Theistic evolutionists who try to blend evolution with the Bible by using the Gap Theory are wrong. If evolution ever occurred, the geological strata would have recorded it. Rather than proving evolution, the fossils in the geological strata prove that Darwinian evolution never happened. True science filters out false theology.

There were and are many well-educated creationist scholars who believe the Restoration Theory. Scofield wasn't the only one. Another creationist who defended the Restoration Theory was Robert B. Thieme, Jr. (1918-2009). He was an excellent Hebrew and Greek scholar, and he believed in the Theory of Restoration. I would not want to be the man who claimed that Thieme wasn't a Hebrew scholar or that he was ignorant of what the Bible teaches. Thieme was one of the great theologians of our time, and I have tremendous respect for his opinions and his scholarly viewpoints. In his book, *Creation, Chaos, and Restoration,*[22] Thieme explained the Hebrew of Genesis One. He showed how Genesis describes the restoration of earth after a period of Divine judgment. Next, I read the writings of Finis Jennings Dake (1902-1987). His notes and comments in *Dake's Annotated Reference Bible*[23] made it very clear that the Restoration Theory was not only permissible, but it was the only theory that explained some troublesome passages in Isaiah, Ezekiel, and Jeremiah. Again, I would not want to accuse Dake of being uneducated or of being an evolutionist. There were other scholars who defended the Restoration Theory as well. Eric Sauer (1931-)wrote *The King Of The Earth*[24] and George H. Pember (1837-1910) wrote *Earth's Earliest Ages.*[25] They both believed that the earth had been destroyed and later restored.

Of course, this doesn't mean I accept all the various teachings of all these men, but at least their motives for believing the Gap Theory wasn't because they felt a need to defend evolution. Many people dismiss the Restoration Theory as a modern attempt to blend the Bible with Darwin's Theory of Evolution. Those who say this have probably never studied the Gap Theory. Bernard Ramm (1916-1992) studied it, and while not agreeing with it himself, he did show that it wasn't a 19th century invention. In his book, *The Christian View of Science and Scripture,*[26] he listed several pre-Darwinian scholars who believed in the Gap Theory. These include Edgar, King of England (10th century A.D.), Simon Episcopius (1583-1643), J. G. Rosenmuller (1776), William

Buckland (1837), J. Pye Smith (1840), and others. Since the Theory of Evolution wasn't yet written, it's hard to imagine these Restored-Earth scholars were trying to blend the Bible with the Theory of Evolution.

As I continued my research, I discovered more scholars who defended the Gap Theory. John Nelson Darby (1800-1882) wrote:[27]

"The passage in Isa. 45:18, 'he created it not it vain (chaotic)', is conclusive that the earth was not created chaotic at first. The earth got into the state of chaos- it may be what destroyed the animals; but we know nothing about it: what I do know by faith is that God created everything."

Darby believed the earth was not created a chaos. Something caused the earth to become one. He also believed that the animals on earth were destroyed during that time. Was Darby unscholarly? Did he not know Hebrew? No, and no! He was a qualified theologian and Hebrew scholar, and he believed that something bad happened between Genesis 1:1 and Genesis 1:2.

Jamieson, Fausset, and Brown echo the thought in their commentary (1871) on Genesis 1:2.[28]

"the earth was without form and void— or in 'confusion and emptiness,' as the words are rendered in Isa. 34:11. This globe, at some undescribed period, having been convulsed and broken up, was a dark and watery waste for ages perhaps, till out of this chaotic state, the present fabric of the world was made to arise."

Alfred Edersheim[29] (1825-1889) made this comment about Genesis 1:1 and Genesis 1:2.

"The first verse in the book of Genesis simply states the general fact, that 'In the beginning' - whenever that may have been – 'God created the heaven and the earth.' Then, in the second verse, we find earth described as it was at the close of the last great revolution, preceding the present state of things: 'And the earth was without form and void; and darkness was upon the face of the deep.' An almost indefinite space of time, and many changes, may therefore have intervened between the creation of heaven and earth, as mentioned in ver. 1, and the chaotic state of our earth, as described in ver. 2."

These scholars agreed that something caused the earth to become a dead and dark watery waste. A. W. Pink[30] (1886-1952), another scholar who believed the Gap Theory, wrote:

"What is found in the remainder of Genesis 1 refers not to the primitive creation but to the *restoration* of that which had fallen into ruins. Genesis 1:1 speaks of the original creation; Genesis 1:2 describes the then condition of the earth six days before Adam was called into existence. To what remote point in time Genesis 1:1 conducts us, or as to how long an interval passed before the earth "*became*" a ruin, we have no means of knowing; but if the surmises of geologists could be conclusively established there would be no conflict at all between the findings of science and the teaching of Scripture. The unknown interval between the first two verses of Genesis 1, is wide enough to embrace all the prehistoric ages which may have elapsed; but all that took place from Genesis 1:3 onwards transpired less than six thousand years ago." (emphasis his)

Lewis Sperry Chafer[31] (1871-1952), the founder and first president of Dallas Theological Seminary, believed in the Gap Theory. I find it difficult to understand how those who hate the Gap Theory can continue to say that qualified biblical scholars don't support it. Chafer was qualified enough to establish and preside over a very prestigious theological seminary. Here is his comment on Ezekiel 28:11-19, a passage that deals with Lucifer's origin and fall:

"*Ezekiel 28:11-19.* A considerable portion of this immediate context is to be taken up verse by verse, but in preparation for that understanding it may be observed that revelation concerning Satan begins with the dateless period between the creation of the heavens and the earth in that perfect form in which they first appeared (Gen. 1:1) and the desolating judgments which ended that period, when the earth became waste and empty (Gen. 1:2; Isa. 24:1; Jer. 4:23-26)."

Gleason Archer[32] (1916-2004), a well-known and highly respected Hebrew scholar, also recognized the possibility of a gap between Genesis 1:1 and Genesis 1:2. This is significant because Archer believed in the Day-Age Theory instead of the Gap Theory. He truly was a "Frank Morison" among the scholars. Although he didn't believe the Ruin-Restoration Theory, he certainly knew it was possible, and he wasn't afraid to express a truth that might be contrary to his own ideas.

"It should be noted in this connection that the verb was in Genesis 1:2 may quite possibly be rendered 'became' and be construed to mean: 'And the earth became formless and void.' Only a cosmic catastrophe could account for the introduction of chaotic confusion into the original perfection of God's creation. **This interpretation certainly seems to be exegetically tenable...**"

Archer also revealed a very important fact about the meaning of *HAYAH*.

"Properly speaking, this verb *hayah* **never has the meaning of static being like the copular verb 'to be.' Its basic notion is that of becoming or emerging as such and such, or of coming into being** . . . Sometimes a distinction is attempted along the following lines: *hayah* means 'become' only when it is followed by the preposition *le*; otherwise there is no explicit idea of becoming. But this distinction will not stand up under analysis. In Genesis 3:20 the proper rendering is: 'And Adam called the name of his wife Eve, because she became the mother of all living.' No *le* follows the verb in this case. So also in Genesis 4:20: 'Jabal became the father of tent dwellers.' **Therefore there can be no grammatical objection raised to translating Genesis 1:2: 'And the earth became a wasteness and desolation.'"**

Who wants to go on record saying that Dr. Archer, a professor of Hebrew at one of America's highly-respected schools of divinity, didn't know how to translate Hebrew? Many Young-Earth and Day-Age creationists think he didn't. They continue preaching the tired claim that *HAYAH* cannot be translated "became." They obviously believe that Archer was unscholarly. They believe the same thing about Hebrew scholar Martin Anstey when he expressed his scholarly opinion in his book, *The Romance of Bible Chronology*[33] (1913).

"The opening verse of Genesis speaks of the Creation of the heavens, and the earth, in the undefined beginning. From this point we may date the origin of the world, but not the origin of man. For the second verse tells of a catastrophe - the earth became a ruin, and a desolation. The Hebrew verb *hayah* (*hayah* = to be) here translated was, signifies not only 'to be' but also 'to become,' 'to take place,' 'to come to pass.' When a Hebrew writer makes a simple affirmation, or merely predicates the existence of anything, the verb *hayah* is **never expressed**. Where it is expressed **it must always be translated by our verb to become**, never by the verb to be, if we desire to convey the exact shade of the meaning of the Original. The words *tohu va-bohu*, translated in the A.V. 'without form and void' and in the R.V. 'waste and void' should be rendered *tohu*, a ruin, and *bohu*, a desolation. They do not represent the state of the heavens and the earth as they were created by God. They represent only the state of the earth as it afterwards became - 'a ruin and a desolation.' This interpretation is confirmed by the words of Isaiah 45:18,

'He created it not *tohu* (a ruin): He formed it to be inhabited (habitable, not desolate).'

This excludes the rendering of Gen. 1:2 in the A.V. and the R.V. as decisively as the Hebrew of Gen. 1:2 requires the rendering of *hayah* by the word 'became' instead of the word 'was,' or better still 'had become,' **the separation of the *Vav* from the verb being the Hebrew method of indicating the pluperfect tense.**"

The *WAW (VAV)* is separated from the *HAYAH* in Genesis 1:2. This means it is in the pluperfect tense. The pluperfect tense designates an action completed before a specific point of past time. Examples are: "I had flown to New York before flying to London." "Billy had eaten three cupcakes when his mother caught him." The pluperfect is the "had" form of a past completed tense verb. You can't combine "had" with "was" as a form of a past completed action. Such a translation would read, "And the earth had was without form, and void." It makes no sense. **The *HAYAH* of Genesis 1:2 is not merely describing a condition; it is describing an action.** "Was" in a static sense doesn't fit!

Two more highly qualified and knowledgeable Hebrew scholars who affirm (2003) that *HAYAH* means "to become" are Warren Baker and Eugene Carpenter:[34]

"*hāyāh*: A verb meaning to exist, to be, to become, to happen, to come to pass, to be done. It is used over 3,500 times in the Old Testament. In the simple stem, the verb often means to become, to take place, to happen. It indicates that something has occurred or come about, such as events that have turned out a certain way (1 Sam. 4:16); something has happened to someone, such as Moses (Ex. 32:1, 23; 2 Kgs. 7:20); or something has occurred just as God said it would (Gen. 1:7, 9). Often a special Hebrew construction using the imperfect form of the verb asserts that something came to pass (cf. Gen. 1:7, 9). Less often, the construction is used with the perfect form of the verb to refer to something coming to pass in the future (Isa. 7:18, 21; Hos. 2:16).

The verb is used to describe something that comes into being or arises. For instance, a great cry arose in Egypt when the firstborn were killed in the tenth plague (Ex. 12:30; cf. Gen. 9:16; Mic. 7:4); and when God commanded light to appear, and it did (Gen. 1:3)."

Charles Andrew Coates[35] (1862-1945) defended the Gap Theory.

"In the beginning God created the heavens and the earth." That is all we get about the original creation. Then in the second verse we find things fallen into a state of ruin. "And the earth was waste and empty, and darkness was on the face of the deep." This was certainly not as it was created…"

William Kelly[36] (1820-1906) also believed the earth had two beginnings.

"...the first verse speaks of an original condition which God was pleased to bring into being; the second, of a desolation afterwards brought in; but how long the first lasted what changes may have intervened, when or by what means the ruin came to pass, is not the subject-matter of the inspired record..."

Arno Clement Gaebelein[37] (1861-1945) wrote about how the earth became waste and void.

"The original earth passed through a great upheaval. A judgment swept over it, which in all probability must have occurred on account of the fall of that mighty creature, Lucifer, who fell by pride and became the devil. The original earth, no doubt, was his habitation and he had authority over it which he still claims as the prince of this world. Luke 4:5-6 shows us this. The earth had become waste and void; chaos and darkness reigned. What that original earth was we do not know, but we know that animal and vegetable life was in existence long before God began to restore the earth."

Frederick William Grant[38] (1834-1902) spoke of the fall of the original earth.

"For plainly the work of the six days begins with this: "God said, 'Let there be light;' and there was light." But as plainly the earth, although waste and desolate, was there before that, not created then. Moreover the words "without form and void," for which "waste and desolate" would be preferable as a reading, imply distinctly a state of ruin, and not of development;... There was, then, a primary creation, afterward a fall;"

Frank Binford Hole[39] (1874-1964) taught that the earth suffered a catastrophe long after its original creation.

"In verse 2 we move from that remote epoch to a time much nearer our own, and we descend, as regards this earth, to a state of very great imperfection. It is found 'without form;' that is, a ruin, a waste: it is also 'void:' that is, empty. Isaiah 45:18 plainly says, 'He created it not in vain, He formed it to be inhabited.' This is very striking, for here again the proper word for creation is used, as in our first verse, and 'in vain' is a translation of the same word as 'without form' in our verse. So we have a definite confirmation of the thought that the state of the earth as in verse 2, was one that supervened, long after the original creation, as the result of some catastrophic event which is not revealed to us."

Albert Barnes[40] (1798-1870) noted the significance of the use of *HAYAH* in Genesis 1:2.

"If the verb had been absent in Hebrew, the sentence would have been still complete, and the meaning as follows: 'And the land was waste and void.' **With the verb present, therefore, it must denote something more.** The verb היה hāyâh 'be' has here, we conceive, the meaning 'become;' and the import of the sentence is this: 'And the land had become waste and void.'"

Louis Ginsberg (1873-1953), noted professor and Jewish historian, wrote about how the ancient Jews actually believed there had been more than one world before the present world.[41]

"Nor is this world inhabited by man the first of things earthly created by God. **He made several worlds before ours**, but He destroyed them all, because He was pleased with none until He created ours."

These scholars haven't hidden their works. This information is readily available to anyone who seeks truth. In spite of this, I can go to any Christian bookstore and purchase books written by Christians who say that *HAYAH* cannot be translated "became." Right now, I can go to dozens of Internet web sites (created and maintained by Christians) that vehemently proclaim the Gap Theory is without merit because there is no justification for translating *HAYAH* as "became." Today, I could go to any of the Christian bookstores in town and find books written by Christian creationists who say there are few scholars who defend the Gap Theory. Some say there are none. My main concern for such untrue proclamations is that unbelievers are offended by Christians who knowingly lie. Why would they believe our claims about Christ if we are shown to be liars? **Would you believe the claims of liars?**

Without Form, and void

When I purchased the book, *Without Form and Void*,[42] by Arthur C. Custance (1910-1985), I found what I consider to be the best source available for determining the meaning of Genesis 1:1-2. More than any other book, his book convinced me of the truth underlying the Restoration Theory. If you want to know what *HAYAH* means in Genesis 1:2, then you need to read *Without Form and Void*. Don't let someone convince you to reject the Restoration Theory if you haven't read this very scholarly work. If someone criticizes the Restoration Theory, ask them, "Have you read Arthur Custance?" I can't encourage you enough to purchase and read it. In fact, if you could afford no other book on the subject, this is the book to buy. Custance was most certainly a Hebrew scholar, and he made no attempt to blend evolution with the Bible. His book explains in detail why the grammar of Genesis 1:2 is best translated "but the earth had become..." Furthermore, he shows that the Restoration Theory was believed long before Darwin. In fact, it predates the Christian Church. Custance explains how the Targum of Onkelos, the earliest Aramaic Old Testament (Dated anywhere from the 2nd century B.C. to the 2nd century A.D.), spoke of the earth being "laid waste" in Genesis 1:2. Custance also reveals that many of the early Church fathers, including Justin Martyr, St. Gregory Nazianzen, Origen, Theodoret, and Augustine believed there was a long gap of time between the creation of the earth and the creation of Adam. (Although not all of them described it as a period of Ruin-Restoration.) He also includes the names of several well-known scholars who wrote and defended some form of the Ruin-Restoration Theory before Darwin wrote about his Theory of Evolution in 1859. These include Simeon ben Jochai (2nd century A.D.), Caedmon (A.D. 650), Alcuin of York (A.D. 735-804), Hugo St. Victor (A.D. 1097-1141), Thomas Aquinas (A.D. 1226-1274), Benedict Pererius (16th century A.D.), Dionysius Petavius (A.D. 1583-1652), Johann August Dathe (A.D. 1731-1791), Thomas Chalmers (A.D. 1780-1847), John Harris (A.D. 1802-1856), J. H. Kurtz (A.D. 1809-1890), and others. In all, he mentions directly or indirectly about eighty scholars who have expressed their belief in some form of a gap of time between Genesis 1:1 and Genesis 1:3. I highly encourage you to purchase and study this book. It can be ordered at:

Doorway Publications
38 Elora Drive Unit 41
Hamilton, Ontario, Canada L9C 7L6

(It is also available on the Internet at http://www.custance.org)

Finally, for those of you who still believe that *HAYAH* cannot be translated "became," I now direct your attention to the footnote on Genesis 1:2 in the *New International Version*.

The NIV translates it as "was," but the translators add this comment: "or possibly *became*." They didn't translate it as "became," but they did say it was possible. I don't think the NIV translators were ignorant of Hebrew grammar rules. I don't think they were trying to defend the Gap Theory either. Yet, if translating *HAYAH* as "became" was impossible, then why did these Hebrew scholars say that it was possible?

It becomes apparent then that it is proper to translate Genesis 1:2 as: "but the earth had become without form, and void." How you translate it depends on more than just the rules of grammar; it depends on what you think the sequence of events was. This in turn depends on what you think God did at the beginning. We find ourselves stuck in a loop. What we think happened in the beginning governs how we translate this passage. How we translate this passage governs what we think happened in the beginning. To translate this passage with the information God intended to reveal requires that we know something about the events and details of the beginning. We need to know the truth in order to know the truth. Since none of us were present at the beginning, our knowledge of the beginning is dependent on God's revelation. The more truth God reveals to us, the more truth we can know. The more truth we know, the better we can translate Genesis 1:1-2. This is why we must not remain fixed in Genesis One. We must see what God has revealed about the creation in other parts of the Bible.

Another Creation Account

My belief in Theistic Evolution was shaken when I was challenged with Scripture indicating that Adam came before Eve. I knew enough about science to know that men couldn't have evolved without women. The Bible said Adam was alone, but that would be evolutionarily impossible. I knew something was wrong with my belief about evolution.

Likewise, my Young-Earth beliefs about creation were shaken when I was challenged with Scripture indicating God did not originally create earth as a dead and dark desolate waste. It is important that you realize my change in creation theories did not come about because of any scientific theory. I did not change my mind because I felt compelled to compromise the Bible with evolution. My change of theories came about because I firmly believe that Scripture cannot contradict Scripture. If God originally created earth a desolate waste, then the Bible contradicts the Bible. The earth was a desolate waste at the beginning of the six days of creation, but it was not created that way.

Isa. 45:18 "For thus saith the LORD that created the heavens; God himself that formed the earth and made it; he hath established it, he created it not in vain, he formed it to be inhabited: I *am* the LORD; and *there is* none else." (KJV)

Isa. 45:18 "For thus says the Lord, who created the heavens (he is God!), who formed the earth and made it (he established it; he did not create a chaos, he formed it to be inhabited!):" (RSV)

Isa. 45:18 "For thus says the Lord, who created the heavens, (He is the God who formed the earth and made it, He established it and did not create it a waste place, But formed it to be inhabited.)" (NASB)

Isa. 45:18 "For thus says the Lord who created the heavens— He is the true God— Who formed the earth and made it— He established it— He created it not a chaos, He formed it for a dwelling place." (S&G)

Isa. 45:18 "For thus says the Lord, who created the heavens (he is God!), who formed the earth and made it (he established

it; he did not create a chaos, he formed it to be inhabited!):"
(NRSV)

Guess what Hebrew word God used when He said He
did not create the earth a chaos? He used *TOHUW*, the same
word He used to describe the earth in Genesis 1:2! God is
giving us another account of earth's creation. As we have
seen, God gave Job some additional details about the creation of the earth. He does the same for Isaiah. He's giving
Isaiah a physical description of the earth at the time He created it. God is telling Isaiah that He did not create it *TOHUW*.
God did not create earth a desolate waste. There are those
who would like you to believe that *TOHUW* is used in the
figurative sense here in Isaiah 45:18 but in the literal sense in
Genesis 1:2. This isn't so. There is nothing in Isaiah 45 indicating that God is speaking figuratively when He describes
the creation of earth. In this chapter, God describes literal
things about Cyrus, Jacob, the heavens, the stars, the sun,
Egypt, Israel, the merchants of Ethiopia, the Sabeans, carved
images, and other physical items. None of these are figurative. God is literal in his descriptions of events, people, and
places. God is literal in his description of earth. It would
be very strange to shift from giving literal descriptions to
giving figurative symbols without any clues. God describes
literal things before verse 18 and literal things after verse 18.
What makes anyone believe He's not being literal in verse
18? The answer is this: If *TOHUW* in Isaiah 45:18 is a literal
description of earth, then the earth was not created a literal
desolation, and that throws a monkey wrench into a lot of
established theology.

The Meaning of *TOHUW* in Isaiah 45:18

I believe Isaiah has given us a description of what the
earth was like when it was first created. However, there can

be two different interpretations of this passage. *TOHUW* can be translated as an OBJECT. Some translations say that God did not create "a chaos," "a waste," or "a desolation." *TOHUW* can also be translated as a PURPOSE. Some translations tell us that God did not create "in vain" or "for naught." Which is correct? Again, I think we have a situation where both interpretations are grammatically correct, but only one interpretation conveys the information God wants us to know. Why is this important to the Gap Theory? Because if *TOHUW* in Isaiah 45:18 describes an object, "a desolate waste," then the earth was not created that way. This would mean that at the end of Genesis 1:1, the earth was not a desolate waste. It would imply that earth was created in a beautiful, well-ordered condition. This would further imply that for it to be *TOHUW* in Genesis 1:2, it must have BECOME a desolate waste. On the other hand, if *TOHUW* in Isaiah 45:18 describes a purpose, then this passage has no bearing on the Gap Theory. It would simply be telling us that God had a purpose for earth regardless of its physical condition when created.

So, *TOHUW* can be a noun meaning "desolate waste," or it can be an adverb meaning "in vain." The scholars have long debated over which meaning of *TOHUW* should be used in Isaiah 45:18. Since they can't agree, and since I'm not a Hebrew scholar, I will not be able to prove my belief. However, I think God has given us a clue. As long as we look only at the word *TOHUW*, I don't think we can determine which is more correct. If we look at *BARA*, the word for create, then I think we can discern the meaning of *TOHUW*. When used in regard to God's activities, *BARA* means "to create." It is also used to describe shaping, making, producing, cutting, clearing, and other ideas, especially when used of man's activities. I want to focus on how *BARA* is used in connection with God's activities. Here are the uses of *BARA* in the *King James Version* that describe God's actions.

I will leave Isaiah 45:18 out of the list so we can look at it last. The words in bold print are translations of the verb *BARA* and the underlined words are the objects of *BARA*. Note that *BARA* always refers to a WHAT, never to a WHY or a HOW. The object of *BARA* is always a noun or pronoun. It is a person, place, or thing. It is never an adverb, adjective, or other modifier. Now, I'm not saying that such a use is impossible in Hebrew, but it is not used that way in the Bible when God does the creating. I think this is God's clue to us. It would seem to me that the best translation of *TOHUW* when used with *BARA* is to translate it as the noun object "a desolate waste," not as the adverb modifier "in vain."

BARA

Gen. 1:1 "In the beginning God **created** the <u>heaven</u> and the <u>earth</u>."

Gen. 1:21 "And God **created** <u>great whales</u>, and every <u>living creature</u> that moveth, which the waters brought forth abundantly, after their kind, and every <u>winged fowl</u> after his kind: and God saw that *it was* good."

Gen. 1:27 "So God **created** <u>man</u> in his *own* image, in the image of God **created** he <u>him</u>; <u>male</u> and <u>female</u> **created** he <u>them</u>."

Gen. 2:3 "And God blessed the seventh day, and sanctified it: because that in it he had rested from all his <u>work</u> which God **created** and made."

Gen. 2:4 "These *are* the generations of the <u>heavens</u> and of the <u>earth</u> when they were **created**, in the day that the LORD God made the earth and the heavens,"

Gen. 5:1 "This *is* the book of the generations of Adam. In the day that God **created** man, in the likeness of God made he him; "

Gen. 5:2 "Male and female **created** he them; and blessed them, and called their name Adam, in the day when they were **created**."

Gen. 6:7 "And the LORD said, I will destroy man whom I have **created** from the face of the earth; both man, and beast, and the creeping thing, and the fowls of the air; for it repenteth me that I have made them."

Exo. 34:10 "And he said, Behold, I make a covenant: before all thy people I will do marvels, such as have not been **done** in all the earth, nor in any nation: and all the people among which thou *art* shall see the work of the LORD: for it *is* a terrible thing that I will do with thee."

Num. 16:30 "But if the LORD **make** a new thing, and the earth open her mouth, and swallow them up, with all that *appertain* unto them, and they go down quick into the pit; then ye shall understand that these men have provoked the LORD."

Deu. 4:32 "For ask now of the days that are past, which were before thee, since the day that God **created** man upon the earth, and *ask* from the one side of heaven unto the other, whether there hath been *any such thing* as this great thing *is,* or hath been heard like it? "

Psa. 51:10 "**Create** in me a clean heart, O God; and renew a right spirit within me."

Psa. 89:12 "The <u>north</u> and the <u>south</u> thou hast **created** <u>them</u>: Tabor and Hermon shall rejoice in thy name."

Psa. 102:18 "This shall be written for the generation to come: and the <u>people</u> which shall be **created** shall praise the LORD."

Psa. 104:30 "Thou sendest forth thy spirit, <u>they</u> are **created**: and thou renewest the face of the earth."

Psa. 148:5 "Let them praise the name of the LORD: for he commanded, and <u>they</u> were **created**."

Eccl. 12:1 "Remember now <u>thy</u> **Creator** in the days of thy youth, while the evil days come not, nor the years draw nigh, when thou shalt say, I have no pleasure in them;"

Isa. 4:5 "And the LORD will **create** upon every dwelling place of mount Zion, and upon her assemblies, a <u>cloud</u> and <u>smoke</u> by day, and the <u>shining of a flaming fire</u> by night: for upon all the glory *shall be* a defence."

Isa. 40:26 "Lift up your eyes on high, and behold who hath **created** <u>these *things*,</u> that bringeth out their host by number: he calleth them all by names by the greatness of his might, for that *he is* strong in power; not one faileth."

Isa. 40:28 "Hast thou not known? hast thou not heard, *that* the everlasting God, the LORD, the **Creator** of <u>the ends of the earth,</u> fainteth not, neither is weary? *there is* no searching of his understanding."

Isa. 41:20 "That they may see, and know, and consider, and understand together, that the hand of the LORD hath done this, and the Holy One of Israel hath **created** <u>it</u>."

Isa. 42:5 "Thus saith God the LORD, he that **created** the <u>heavens</u>, and stretched them out; he that spread forth the earth, and that which cometh out of it; he that giveth breath unto the people upon it, and spirit to them that walk therein:"

Isa. 43:1 "But now thus saith the LORD that **created** <u>thee</u>, O Jacob, and he that formed thee, O Israel, Fear not: for I have redeemed thee, I have called *thee* by thy name; thou *art* mine."

Isa. 43:7 "*Even* every one that is called by my name: for I have **created** <u>him</u> for my glory, I have formed him; yea, I have made him."

Isa. 43:15 "I *am* the LORD, your Holy One, the **creator** of <u>Israel</u>, your King."

Isa. 45:7 "I form the light, and **create** <u>darkness</u>: I make peace, and **create** <u>evil</u>: I the LORD do all these *things*."

Isa. 45:8 "Drop down, ye heavens, from above, and let the skies pour down righteousness: let the earth open, and let them bring forth salvation, and let righteousness spring up together; I the LORD have **created** <u>it</u>."

Isa. 45:12 "I have made the earth, and **created** <u>man</u> upon it: I, *even* my hands, have stretched out the heavens, and all their host have I commanded."

Isa. 48:7 "<u>They</u> are **created** now, and not from the beginning; even before the day when thou heardest them not; lest thou shouldest say, Behold, I knew them."

Isa. 54:16 "Behold, I have **created** the <u>smith</u> that bloweth the coals in the fire, and that bringeth forth an instrument for his work; and I have **created** the <u>waster</u> to destroy."

Isa. 57:19 "I **create** the <u>fruit of the lips</u>; Peace, peace to *him that is* far off, and to *him that is* near, saith the LORD; and I will heal him."

Isa. 65:17 "For, behold, I **create** <u>new heavens</u> and a <u>new earth</u>: and the former shall not be remembered, nor come into mind."

Isa. 65:18 "But be ye glad and rejoice for ever *in that* which I **create**: for, behold, I **create** <u>Jerusalem</u> a rejoicing, and her people a joy."

Jer. 31:22 "How long wilt thou go about, O thou backsliding daughter? for the LORD hath **created** <u>a new thing</u> in the earth, A woman shall compass a man."

Eze. 21:30 "Shall I cause *it* to return into his sheath? I will judge thee in the place where <u>thou</u> wast **created**, in the land of thy nativity."

Eze. 28:13 "Thou hast been in Eden the garden of God; every precious stone *was* thy covering, the sardius, topaz, and the diamond, the beryl, the onyx, and the jasper, the sapphire, the emerald, and the carbuncle, and gold: the workmanship of thy tabrets and of thy pipes was prepared in thee in the day that <u>thou</u> wast **created**."

Eze. 28:15 "Thou *wast* perfect in thy ways from the day that <u>thou</u> wast **created**, till iniquity was found in thee."

Amos 4:13 "For, lo, he that formeth the mountains, and **createth** the <u>wind</u>, and declareth unto man what *is* his thought, that maketh the morning darkness, and treadeth upon the high places of the earth, The LORD, The God of hosts, *is* his name."

Mal. 2:10 "Have we not all one father? hath not one God **created** <u>us</u>? why do we deal treacherously every man against his brother, by profaning the covenant of our fathers?"

Each time *BARA* is used in these passages, it refers to some THING being created, some object. Great whales are created. Winged fowls are created. Man is created. The universe is created. Jerusalem is created. A new heavens and a new earth are created. A new heart is created. These are **objects** of *BARA* and not **modifiers** of *BARA* when used of God's creations. Let me clarify what this means by using an example in English.

1) Jim built the house.   2) Jim built skillfully.

In the first sentence, "the house" describes WHAT Jim built. In the second sentence, "skillfully" describes HOW Jim built. The first sentence contains an OBJECT of the verb, while the second sentence contains a MODIFIER of the verb. We could describe other modifiers of Jim's work. We could say, "Jim built joyfully," or "Jim built quickly," or "Jim built in vain," or any other number of things that describe how, or why, or when Jim built. But these things are never WHAT Jim built. Only an OBJECT of the verb will tell us WHAT Jim built. The subject and the verb are identical, but the information gleaned from the sentences is different. One has an object; the other has a modifier. Now let's look at Isaiah 45:18 again. We have two choices.

Isa. 45:18 "For thus saith the LORD that **created** the heavens; God himself that formed the earth and made it; he hath established it, he **created** it not in vain, he formed it to be inhabited: I *am* the LORD; and *there is* none else." (KJV)

Isa. 45:18 "For thus says the LORD, who **created** the heavens (he is God!), who formed the earth and made it (he established it; he did not **create** it a chaos, he formed it to be inhabited!): I am the LORD, and there is no other." (NRSV)

Throughout the rest of Old Testament, when *BARA* is used to describe God's creations; it describes God creating an object or objects. The logical choice for Isaiah 45:18 is to accept *TOHUW* as a physical object, the object of *BARA*. This is the way all the other words are used in connection with *BARA* when God creates. This is the way *BARA* is translated even in the first part of Isaiah 45:18, "the LORD, who **created** the heavens." This first usage should set the context. Isaiah is talking about the creation of the heavens and the earth. He is giving us another account of creation. Isaiah's contemporaries would have immediately thought of Genesis when they read his words. That seems to be Isaiah's intent. Isaiah seems to be telling us that God did not create a *TOHUW*, a desolate waste, when He created the earth. Now, it is true that Isaiah includes a purpose statement (a modifier of the verb) later in the sentence. He says, "He formed it to be inhabited," but this purpose statement does not include *BARA*. Instead of *BARA*, Isaiah says, "He formed (*YATSAR*) it to be inhabited." *YATSAR* is not the same as *BARA*. Neither is the word *ASAH* when he says, "... and made (*ASAH*) it." All three words are used in this verse. Although these words are sometimes used interchangeably, God is making a distinction between *BARA*, *YATSAR*, and *ASAH* in Isaiah 45:18. The first *BARA* in this verse tells us WHAT God created. The second *BARA* in this verse also tells us WHAT God created. Isaiah tells us that God created an earth that wasn't a chaos.

Isaiah 45:18 and Genesis 1:1 both describe the creation. Both passages use the word *TOHUW*; both passages talk about the heavens; both talk about the earth. They both show WHAT God created. All these things are literal. The creation isn't figurative. The heavens and earth aren't figurative. God isn't figurative. Isaiah and Moses both talk about the same creation. You can't let one of these passages be figurative without both passages being figurative. If one is literal, the other is literal. They both describe the same God, the same creation, the same earth. They both use *BARA* and they both use *TOHUW*. The only difference is that in Isaiah, God tells us that He did not create earth *TOHUW*, but in Genesis, God tells us that earth either "was" or "became" *TOHUW*. Both passages describe the creation of the heavens. Both passages describe the creation of the earth. According to Isaiah 45:18, at its creation, earth was not a desolate waste, but at Genesis 1:2, it was a desolate waste. Genesis 1:2 must therefore describe what the earth had become sometime after its original, non-*TOHUW* creation in Genesis 1:1. The only way that I can see how both passages could be true is if Genesis 1:2 reads: "… but the earth had become without form, and void."

Before I leave the subject of *BARA* and *ASAH* let me clarify something. Some Young-Earth scholars insist that *BARA* and *ASAH* are always synonymous. They say that there are no distinctions between these two words and they can be used interchangeably throughout Scripture. They say this because they point to Exodus 20:11 as proof that God created the universe only a few thousand years ago.

Exo. 20:11 "For *in* six days the LORD **made** heaven and earth, the sea, and all that in them *is,* and rested the seventh day: wherefore the LORD blessed the sabbath day, and hallowed it." (KJV)

The word used in Exodus 20:11 is *ASAH*. The Lord made (*ASAH*) the heavens and the earth. So, they argue that there is no difference between God making (*ASAH*) the heavens and the earth in Exodus 20:11 and God creating (*BARA*) the heavens and the earth in Genesis 1:1. In their way of thinking, this eliminates the possibility that God created (either *BARA* or *ASAH*) anything before the six days. This would disprove the Gap Theory, but I think this argument fails. Exodus 20:11 describes what God did **during the six days** of Genesis 1:3-31, not what He did in Genesis 1:1. *ASAH* is a better word than *BARA* to describe God's works over the entire six day period because it is more inclusive than *BARA*. In fact, God Himself used *ASAH* to summarize what He did <u>during</u> the six days.

Gen. 1:31 "And God saw every thing that he had **made**, and, behold, *it was* very good. And the evening and the morning were the sixth day." (KJV)

This shows that *ASAH* can include "creating," "making," and "forming" because God did all three of those things during the six days. The implication of Exodus 20:11 seems to me that God *ex nihilo* created some things. Then He made other things out of the preexisting things He had created. "Making" would be a better word to describe all of these actions because "making" can also include the idea of "creating." Over those six days He **created** some things, **made** other things, and **formed** yet other things.

Isa. 45:18 "For thus saith the LORD that **created** the heavens; God himself that **formed** the earth and **made** it; he hath established it, he **created** it not in vain, he **formed** it to be inhabited: I *am* the LORD; and *there is* none else." (KJV)

Since this describes "creating," "making," and "forming," I believe we are looking at more than just the *ex nihilo* creation of Genesis 1:1. Genesis 1:1 describes the *ex nihilo* creation. I believe Genesis 1:3-31 describes what God did after, long after, the original *ex nihilo* creation. The creative acts of God during the six days included *ex nihilo* creations, but they were not the same creation as Genesis 1:1. I believe the six days were a restoration, or at least a partial restoration, of whatever it was that Lucifer and his angels corrupted. I think this included both the heavens and the earth since all creation groans.

Rom. 8:22 "For we know that the whole creation groans and suffers the pains of childbirth together until now." (NASB)

All three words, *BARA*, *ASAH*, and *YATSAR* are used in Isaiah 45:18, but this doesn't mean the words are equivalent. There are slight differences of meanings in these words. If they don't mean slightly different things, then Isaiah 43:7 makes no sense.

Isa. 43:7 "Everyone who is called by My name, And whom I have **created** for My glory, Whom I have **formed**, even whom I have **made**." (NASB)

Everyone called by God has been created, (*BARA*) formed, (*YATSAR*) and made (*ASAH*) by God. (I find it fascinating that God uses the same three words for creating man that He uses for creating the universe. There seems to be a parallel between these two creations.) If all these words mean the same, then no one was ever physically conceived, physically developed *en utero*, and then physically born. If all these words mean the same thing, then all of us were created *ex nihilo*. If you think you were created *ex nihilo*, go ask your mother. She'll tell you otherwise! In spite of

202

what some Young-Earth creationists say, these words don't always mean the same thing. Now, while I have no doubt that these two words CAN be used interchangeably in certain places in Scripture, I don't believe they are interchangeable throughout Scripture. In fact, this Young-Earth view is a dangerous idea. If *ASAH* always means the exact same thing as *BARA*, and if they are completely interchangeable, then we can determine HOW (the mechanics of) God created (*BARA*) the universe simply by looking at the meaning of *ASAH*.

Gen. 1:11 "And God said, Let the earth bring forth grass, the herb yielding seed, *and* the fruit tree **yielding** fruit after his kind, whose seed *is* in itself, upon the earth: and it was so." (KJV)

ASAH is used to describe fruit trees making fruit. We know how this happens. Fruit trees take in preexisting water, nutrients, and minerals from the soil. They take in preexisting carbon dioxide from the air. They harness preexisting energy from sunlight. From all this preexisting matter and energy, they make (*ASAH*) new fruit. Now, if *ASAH* is always identical to *BARA*, then we would have to agree that God used preexisting matter and energy to create (*BARA*) the universe. This argument by Young-Earth creationists winds up disproving the very method of creation (*ex nihilo*) they seek to defend. This idea creates a bigger Bible problem than it solves.

Chapter Eight:

A Closer Look at the Six Days

—⚛—

D on't you love those quizzes that give you a pattern to complete like, "What's the next number in this series 2... 4... 8... 16... 32... ___?" Well, we're going to do the same thing with the first chapter in Genesis. Each of the six days begins with a Divine decree of, "And God said, Let..." Each day ends with, "and the evening and the morning were..." The Bible reveals when the days began and when the days ended. God makes the beginning and ending of each day very obvious. God tells us what He created on each particular day. He tells us WHAT was created WHEN. So, let's take a little quiz to see if we know WHAT was created WHEN. We will start at day six and work back to day one.

The Six Days of Creation

DAY	BEGAN	ENDED
6	Verse 24	Verse 31
	And God said, Let the earth bring forth the living creature after his kind, cattle, and creeping thing, and beast of the earth after his kind: and it was so.	And the evening and the morning were the sixth day.
5	Verse 20	Verse 23
	And God said, Let the waters bring forth abundantly the moving creature that hath life, and fowl *that* may fly above the earth in the open firmament of heaven.	And the evening and the morning were the fifth day.
4	Verse 14	Verse 19
	And God said, Let there be lights in the firmament of the heaven to divide the day from the night; and let them be for signs, and for seasons, and for days, and years:	And the evening and the morning were the fourth day.

DAY	BEGAN	ENDED
3	Verse 9	Verse 13
	<u>And God said, Let</u> the waters under the heaven be gathered together unto one place, and let the dry *land* appear: and it was so.	<u>And the evening and the morning were</u> the third day.
2	Verse 6	Verse 8
	<u>And God said, Let</u> there be a firma-ment in the midst of the waters, and let it divide the waters from the waters.	<u>And the evening and the morning were</u> the second day.
1	Verse ????	Verse 5
	???????????????????? ???????????????????? ???????????????????? ???????????????????? ????????????????????	<u>And the evening and the morning were</u> the first day.

Now answer this question:
When did the first day of creation begin?

Do you need a clue? Okay, look for the verse BEFORE VERSE FIVE that says, "And God said, Let…" Look closely; don't be fooled by any preconceived ideas! The answer is verse three. Verse three says, "And God said, Let there be light…" The first day of creation begins at Genesis 1:3, not at Genesis 1:1. Genesis 1:1-2 describe events BEFORE the first day of creation. This has always been a problem for those who reject the Gap Theory. This is why many of them want to make Genesis 1:1 a title or a summary instead of an act of creation. If Genesis 1:1 is an act of creation, and it must be because it says, "God created," then God created the heavens and the earth (the universe) before the first day of creation, before verse three. So what did He create when it says He created the heavens? If it was the stars and galaxies, then what did He create on day four? God reveals two different creations. Genesis 1:1 is the initial creation while Genesis 1:3-31 is a second creation. The standard objection to this is that Genesis 1:1-2 is a summary of what God does during the six days of creation. Such an interpretation creates a contradiction. If Genesis 1:1-2 is a summary of what God does during the six days, then the summary is erroneous. According to Genesis 1:31, at the end of the six days, the earth is inhabited by life. Yet, Genesis 1:2 ends with the earth dead and desolate. At the end of Genesis 1:31, there is light. At the end of Genesis 1:2, the earth is in darkness. At the end of Genesis 1:31, the earth is "very good." At the end of Genesis 1:2, the earth is "without form, and void." The conditions don't match. This is why I reject the idea that Genesis 1:1-2 is a summary. It would be hard to imagine that Genesis 1:1-2 summarize what God did during the creation week when the summary makes such a big mistake. Did God fail to realize this when He inspired Moses to write? Genesis 1:1-2 can't be a summary of Genesis 1:3-31 because they start and end with different conditions. Therefore, **Genesis 1:1-2 must have occurred before the events of Genesis 1:3-31**. The original creation came before the six days. Genesis 1:1 is a creative act all by itself. Since Genesis 1:1 is a creative act and not a summary, then God created the heavens and the earth before day one of the Creation Week. God reveals two separate acts of creation. The first is Genesis 1:1 and the second is Genesis 1:3-31. There were two different beginnings

Let's look at this in more detail. If there was only one beginning, then there are some contradictions. What does it mean when God created the "heavens" in Genesis 1:1? What did He create if not the things that are in the heavens? It cannot mean just empty space because in the very same sentence He said He created the earth. So, if He created the heavens, then He **CREATED** (*BARA*) the galaxies, the quasars, the sun, the moon, and the stars before the first day. What then does it mean when it says He **MADE** (*ASAH*) the sun, moon, and stars on the fourth day in Genesis 1:16? Genesis 1:1 and Genesis 1:16 can't be referring to the same events. God couldn't have created the sun and the stars and the moon on day four if He had already created them before the first day. "Made" in Genesis 1:16 must mean something different than "create" in Genesis 1:1.

If there were three days and three nights before day four, what was the source of light? Some Young-Earth creationists believe there was a glowing ball of matter somewhere out in space, but it wasn't the sun. Where they find that in the Bible is beyond me. It doesn't fit with any known scientific discoveries either. Other Young-Earth creationists say that God Himself was the source. "God is light," they will quote. (1 John 1:5) While it is true that God is light, the text doesn't say that God was the source of that light. That is an assumption, not a revelation. If God was the source of that light, then what happened to God during the night? Did He quit being light? Did His attributes change? As you carefully read the text, you will see that God reveals no distinctions between the source of light for first three days and the last three days. God doesn't say that He was that light for three days and then the sun became that light after that. That's what Young-Earthers want you to believe, but no Bible verse states that. There is no biblical indication that the source of light for days one through three differs from the source of light for days four through six. Both of these light sources divided the day from the night in the exact same way for the exact same duration. This would imply that the sun was the source of light for all six days. This would further imply that the sun was created before the fourth day. That agrees with Genesis 1:1, but it appears to contradict Genesis 1:14-18.

Gen. 1:14-19 "And God said, Let there be lights in the firmament of the heaven to divide the day from the night; and let them be for signs, and for seasons, and for days, and years: *{15}* And let them be for lights in the firmament of the heaven to give light upon the earth: and it was so. *{16}* And God made two great lights; the greater light to rule the day, and the lesser light to rule the night: *he made* the stars also. *{17}* And God set them in the firmament of the heaven to give light upon the earth, *{18}* And to rule over the day and over the night, and to divide the light from the darkness: and God saw that *it was* good. *{19}* And the evening and the morning were the fourth day." (KJV)

Those who adhere to the Day-Age Theory can't explain this. The sun couldn't have come billions of years after the earth. It couldn't have been created billions of years after grass and trees and seed bearing plants. Something seems wrong; something is in the wrong order. The Young-Earth Theory has a different explanation: The sun, moon, and stars weren't created until four days after the creation of the earth. But, as we have seen, this raises the question about what it means when it says the "heavens" were created in Genesis 1:1. It raises the question of, "Where is the scientific evidence?" It also raises a question about the purpose or FUNCTION of the light. Genesis 1:14 says the light of day four was, "to divide the day from the night," but the light called into existence on day one (Genesis 1:3) had

ALREADY been dividing the day from the night. This wasn't new. Days one through three had experienced this same division. They already had three days and three nights; three evenings and three mornings. The division of day and night on day four was not a new division. It was not a new function. It was already happening. That light of day one was already dividing the day from the night. The light of day four was not a new creation. If you were standing on the earth during the first three days, you would have seen three periods of dark and three periods of light. Light and dark were already divided. The light and dark periods of days one through three were literal twenty-four hour days. The light and dark periods of days four through six were literal twenty-four hour days. The light of days one through three produced identical twenty-four hour days as the light of days four through six. Genesis 1:14 doesn't hint of a change in the source of light. Genesis 1:14 doesn't say the light on day four had a different duration. Genesis 1:14 doesn't say the light on day four had a different function. Rather, it was a continuation of the same source with the same duration and the same function. The light of days one through three produced the same duration of day and night that the sun produced in days four through six. That's why I think it was the sun. The Bible gives no indication there was a switch in the source of the light that divided the days from the nights. The Bible doesn't change the function of the light in Genesis 1:14. What is different is this: If you look carefully at the passage, you will see there are some NEW FUNCTIONS added to the LIGHTS on day four. The LIGHT of Genesis 1:3 divided the day from the night. In addition to dividing the day from the night, the LIGHTS of Genesis 1:14 were going to be for "signs," for "seasons," for "days," and for "years." Note that it has shifted from "LIGHT" to "LIGHTS." We now include the moon and the stars and we now have four new functions: signs, seasons, days, and years. What are these new functions?

What-If?

Before I talk about these new functions, I want to talk about the "What-If Game." The reason I'm doing this is because I am going to play the "What-If Game" too. I'm going to propose some explanations for what God was doing with the LIGHT and what God was doing with the LIGHTS. The things I am going to propose are not things the Bible directly says happened. They are things that I think fit best with what the Bible does directly say happened. There seems to be some confusion, so everybody proposes "What-If" scenarios. The Young-Earth creationists have two "What-If" proposals. First, "What if God was that source of light?" Second, "What if there was a glowing ball of matter in space?" The Day-Age creationists have only one "What-If" proposal. "What if the days were ages?" Actually, the Day-Age proposal doesn't help at all. How could there be plant life on earth for a billion years but no sun? Anyway, everyone else makes "What-If" proposals, so I feel I should be allowed to do it as well. However, my "What-If" proposals will not be like their "What-If" proposals. Mine will be scientifically and/or biblically plausible. I won't resort to things such as the speed of light being different. Young-Earthers propose things that are highly improbable, and certainly not observable. Their "What-If" list is quite illogical. What if there were no Laws of Thermodynamics? What if light traveled through Riemannian Space? What if tremendous earthquakes caused the earth's crust to create deep depressions at the mouth of every major river? What if gigantic, cataclysmic waves of water, moving hundreds of miles per hour, were powerful enough to rip out forests, but so calm at the same time as to allow pollen to settle gently to the bottom undisturbed? My "What-If" proposals, on the other hand, are things that God has done elsewhere in the Bible or else things that are scientifically observable and probable. Mine are at least believable. That doesn't mean these things ARE what God did, but it certainly means these things are something God WOULD do, COULD do, or HAS done. The things I propose will have scientific or Scriptural precedents.

Lights in the Sky

The Bible says God spoke light into being on day one, (Gen. 1:3) and that this light divided the day and the night. The Bible also says that God made lights on day four. (Gen. 1:14-16) In neither place does the Bible use the word for "create." The most common word used when God creates something new is *BARA*, and God doesn't use *BARA* in Genesis 1:3. A more accurate translation of Genesis 1:3 is, "And God said, 'Let light **come to be**,' and light came to be." The word used for "Let there be" is our old dynamic friend *HAYAH*. God spoke light into being. He doesn't tell us how it happened, but it wasn't necessarily a new creation. God also uses *HAYAH* in Genesis 1:14 when He says, "Let there be lights in the firmament." Continuing this pattern, God doesn't use *BARA* when He describes what He did with the lights in Genesis 1:16. Instead of "create" two great lights, the Bible says He "made" (*ASAH*) two great lights.

Gen. 1:14-18 "And God said, **Let there be** lights in the firmament of the heaven to divide the day from the night; and let them be for signs, and for seasons, and for days, and years: *{15}* And let them be for lights in the firmament of the heaven to give light upon the earth: and it was so. *{16}* And God **made** two great lights; the greater light to rule the day, and the lesser light to rule the night: *he made* the stars also. *{17}* And God set them in the firmament of the heaven to give light upon the earth, {*18*} And to rule over the day and over the night, and to divide the light from the darkness: and God saw that *it was* good." (KJV)

Now, there have been as many disputes over what *BARA* and *ASAH* mean as there have been over what *YOWM* means. Yes, *BARA* and *ASAH* sometimes can be used to mean the same thing. Usually there is a distinction, even though this distinction is often overlooked. One thing is for certain, *BARA* and *ASAH* are not absolutely synonymous. It seems to me to be quite logical to expect the use of different words if there were different beginnings. God created (*BARA*) the heavens and the earth in Genesis 1:1. This absolutely means to create. I believe this included the sun. In Genesis 1:3 and Genesis 1:14, however, it is not a new creation but a command to let light come to be (*HAYAH*). In other words, God called the light to shine where it had been dark.

Whoa, you may immediately ask the question, "If the sun was already in existence, how could the earth be dark?" Well, the Gap Theory proposes some "What-If" explanations too. What if the darkness that enveloped the earth was a special kind of darkness? The Gap Theory holds that the sun had already been created, yet the earth was dark at the beginning of day one. How do I explain this? I don't believe the earth was dark because the sun hadn't yet been created. I believe the earth was dark because it was being judged? A supernatural darkness accompanied God's judgment. Is there a biblical precedent for this? Yes, do you remember what happened between the sixth hour and the ninth hour of Christ's crucifixion? Darkness covered the land (Matthew 27:45). Why did darkness cover the land? Jesus was being judged for our sins. This wasn't an eclipse. The Passover generally falls on the first full moon after the Spring Equinox, and it's impossible for a solar eclipse to happen during a full moon. This was a special kind of darkness. God caused light to cease being. Light no longer was. God made a supernatural darkness on the day Christ was judged for our sins. Darkness is a picture of judgment. Over and over again in Scripture, darkness is connected with God's judgment. What was the ninth plague on Egypt? God brought about a supernatural darkness through Moses because of Egypt's sin.

Exo. 10:22 "And Moses stretched forth his hand toward heaven; and there was a thick darkness in all the land of Egypt three days:" (KJV)

This wasn't an eclipse either. Eclipses don't last three days. This was darkness so thick it could be felt. This was a supernatural darkness. It was dark in Egypt, but it was still light in Goshen where the Jews lived. Why wasn't it dark where the Jews lived? They weren't being judged. Darkness fell on Egypt because Egypt was being judged. The Gap Theory proposes that the earth was dark before day one because it was being judged. Is this impossible? Do you think God would never darken the earth because of judgment?

Isa. 13:9-11 "Behold, the day of the LORD cometh, cruel both with wrath and fierce anger, to lay the land desolate: and he shall destroy the sinners thereof out of it. {10} For the stars of heaven and the constellations thereof shall not give their light: the sun shall be darkened in his going forth, and the moon shall not cause her light to shine. {11} And I will punish the world for *their* evil, and the wicked for their iniquity; and I will cause the arrogancy of the proud to cease, and will lay low the haughtiness of the terrible." (KJV)

The sun will be darkened and the moon and stars will no longer give light. The earth will be dark at the final judgment. Therefore, the idea that God's judgment can cause the earth to be dark is not without Scriptural support. It isn't a biblical impossibility. Yes, this is a "What-If" proposal, but it is a "What-If" proposal that has credibility.

The Gap Theory proposes that the sun and moon and stars were created before Genesis 1:3. The Gap Theory proposes that the earth was dark because God was judging it. Since Adam wasn't created yet, the most likely explanation based on biblical revelation is that God was judging the earth for the sins of Lucifer and his fallen angels. God judged the earth and made it desolate, dead, and dark because of Lucifer's sin. God then called the light of the sun to come back into existence on day one of the Restoration. He didn't create a new source of light. The sun was already there, only now He allowed its light to "be" once again. That light began dividing the days from the nights. Once again there were evenings and mornings, light and dark, but things weren't completely back to normal. The sun still wasn't directly visible and the stars and the moon still could not be seen. How could this be? Here is another "What-If" proposal of the Gap Theory. What if it was so cloudy that the sun, moon, and stars couldn't be seen?

Gen. 1:6-8 "And God said, Let there be a firmament in the midst of the waters, and let it divide the waters from the waters. {7} And God made the firmament, and divided the waters which *were* under the firmament from the waters which *were* above the firmament: and it was so. {8} And God called the firmament Heaven. And the evening and the morning were the second day. (KJV)

The firmament is the atmosphere. The water under the firmament was the ocean. The waters above the firmament were the clouds. What if by the end of the second day, the clouds were so thick that the sun, moon, and stars were not visible even though their light was still present? Is this "What-If" proposal credible? Of course it is. How many times have you gone outside at night to look at the stars and the moon only to discover that it was too cloudy to see them? How many dreary, gray, overcast days have you experienced when you couldn't see the sun? This is no magical "What-If" proposal. The Gap Theory doesn't require the Laws of Physics to be altered or suspended. According to the Gap Theory, it was so cloudy during the first three days that the sun, moon, and stars weren't visible. This happens all the time. There are many days and nights in which the heavenly bodies can't be seen. This "What-If" doesn't require changes to the Laws of Physics or a unique interpretation of the Bible.

The Canopy Theory

If you are familiar with all the different theories and sub-theories, you may think I am now supporting the Canopy Theory. I am not. The Canopy Theory is the idea that all, or almost all, of the water now in our oceans was locked up in a great, watery canopy that covered the earth before day four. Some think it even continued up until the Great Flood. They think this canopy was the source of the water for The Flood. This idea is unworkable. Yes, the atmosphere can hold water, but there is a limit to how much it can hold. The amount of water it can hold depends on its temperature. There is only so much water that air can hold at any given temperature. Very cold air can hold very little water. The air at the North and South Poles is actually very dry. To hold the oceans, the temperature of the atmosphere would have had to be at or near the boiling point. This wouldn't be "good" for the plants God created on day three. In addition, if the waters of the oceans were suspended in the atmosphere, the weight of the water would create an atmospheric pressure of thousands of pounds per square inch. Nothing could live in such pressure. No, these waters above the firmament didn't create a canopy, but they did make it so cloudy that the sun, moon, and stars were not visible.

Rather than an unscientific canopy, here is what I think happened. On day one, God commanded the light of the sun to be visible again. This ended the supernatural darkness that had covered the earth. I believe the earth had become so cold during this dark period that all (or most) of the water had precipitated out of the atmosphere. All of the water was below the firmament, so there were no clouds. Although the earth was covered with water, it doesn't mean that all the water was liquid. I believe that oceans covered part of the earth and ice and snow covered the rest. This isn't scientifically impossible. Today, virtually all of the dry land of Antarctica is under ice and snow and is not visible. If it had been as cold as I suspect it was, then the earth was experiencing a super ice-age. The entire earth was under water, ice, and snow. No dry land was visible. The newly restored light from the sun began warming the earth. In addition, the Spirit of God "hovered" over the face of the waters. The word used in Genesis 1:2 for what the Spirit was doing is *RACHAPH* and it is also used for an eagle brooding her eggs. Once this warming process started, the waters began evaporating and clouds formed on day two. There was now water (clouds) above the firmament (sky) as well as water (ocean) below it.

Gen. 1:7 "And God made the expanse, and separated the waters which were below the expanse from the waters which were above the expanse; and it was so." (NASB)

The clouds were so thick that the sun, moon, and stars weren't visible. Then on day three, the warmth of the light and the warmth created by the Holy Spirit caused a great thaw. The ice and snow that had hidden the dry land began to melt. This water flowed into the ocean basins and the dry land finally appeared.

Gen. 1:9 "Then God said, 'Let the waters below the heavens be gathered into one place, and let the dry land appear'; and it was so.' (NASB)

New Functions

At this point there was dry land and light, but it was still too cloudy to see the sun, the moon, and the stars. On day four, God made these heavenly bodies appear in the sky. These lights weren't created (*BARA*) on day four, but they were made (*ASAH*) on day four. They were made in the sense that God now produced lights in the sky. *ASAH* conveys the idea of making, doing, establishing, preparing, providing, producing, bringing forth, or causing. It doesn't necessarily imply creating. The same word is used when Adam and Eve made coverings for themselves when they sewed fig leaves together. They didn't create anything. They made a covering so their nakedness couldn't be seen. In this case, God removed a covering so the lights in the sky could be seen. By making the sky clear, God **made** the sun, the moon, and the stars appear. God caused these lights to be seen in the sky. They needed to be seen because God was going to give them some new functions. Besides dividing the day from the night, these functions were for "signs," for "seasons," for "days," and for "years." God intended for man to understand the significance of these new functions. If the sun, moon, and stars were not visible, these new functions wouldn't be understood. Let's look at these functions.

Years

There are actually two types of years. There is the solar year and the lunar year. The solar year is the time it takes for the earth to make one revolution around the sun. From the perspective of the earth, this is seen as the sun moving in its relative position in the sky with respect to the horizon. During the summer, the sun appears higher in the sky and in the winter, it is lower. The lunar year is a little more difficult and can consist of twelve or thirteen cycles of the phases of the moon. Many cultures used lunar calendars. Israel used a complicated lunar/solar calendar. Anyway, God ascribed the function of measuring the years to the newly visible heavenly bodies. God wanted man to be able to measure the passing of years.

Seasons

The beginning and ending of each of the four seasons can be determined by position of the sun and the stars. The position of the stars is important because the earth's revolution around the sun is not exactly 365 days. It's a little under 365 ¼ days. If we relied on the sun alone, we would soon find that the seasons no longer matched our calendar. This could be disastrous in an agricultural society. That's why we add a leap day in February every four years. Even this is not perfectly accurate so from time to time there are additional corrections to the solar calendar. It is the visibility of the stars that allowed man to calculate these corrections. Nowadays, scientists use atomic clocks, but historically astronomers used the position of the stars to do this.

Days

This is not the same as dividing the day from the night. Instead, this was the function of being able to determine when special days, holy days, feast days, etc. were to occur each year. As I mentioned, the Passover came on a full moon after the Spring Equinox. If the sun and moon were not visible, it would have been impossible for primitive man (man without modern time-keeping equipment) to determine the day of the Passover.

Signs

This apparently refers to omens or portents. I don't know exactly what this entailed, but it seems as if the heavenly bodies could be used as signs of special events. It's possible this meant the calculations of solar and lunar eclipses, but that may not be all. Some people speculate that the Gospel of Jesus Christ is symbolically spelled out by the signs of the Zodiac. That sounds weird to me, and it probably is, but I must admit that God intended the visible heavenly bodies to tell us something. What could the heavenly bodies tell us? I think they were meant to tell us something about God. This is what signs were meant to do. Throughout the Bible when God used signs, it was a way of revealing something about Himself to fallen men. I think He meant to do the same thing here.

Set in the Firmament

God set these lights in the firmament according to Genesis 1:17. The word "set" is *NATHAN*. The image that comes to mind is that God creates the stars and then hangs them in space somewhere. This is very picturesque, but it isn't what *NATHAN* means. It doesn't mean to create. Instead, it conveys the idea of giving, presenting, or bestowing. It is used to appoint, to assign, or to delegate. If a king were to appoint an ambassador to a foreign nation, this is the word that would be used. The king doesn't create the ambassador, but he does set him up with a special mission or purpose. Genesis 1:17 doesn't say God created the lights at this point, but it does say He gave us visible lights in the sky for a special purpose. What is this purpose?

The Invisible Attributes of God

What does the visibility of the heavenly bodies have to do with creation? To be honest, I don't know if I have the answer, but since God gave (*NATHAN*) the heavenly bodies these functions, I'm sure He had a good reason. I'll play the "What-If" Game again. **What if God wanted to reveal something about Himself when we look at the sky?**

Rom. 1:20 "For since the creation of the world His invisible attributes, His eternal power and divine nature, have been clearly seen, being understood through what has been made, so that they are without excuse." (NASB)

The visible reveals the invisible. The visible heavenly bodies reveal God's invisible attributes. What do they reveal? For starters, the precision of the movements of the heavenly bodies has always revealed that He is a God of supreme power and order. These are two of His invisible attributes. Gazing at the stars has always made man wonder about the origin of life and the beginning of the universe. They seem to cry out for an eternal, intelligent, supernatural Creator and Life-Giver. These are five more of His invisible attributes. The sun is always where it should be. The moon never fails to be on time. The stars are so constant in their movements that mariners and travelers can determine their latitude and longitude by measuring the position of the stars in the night sky. The celestial bodies show that God is dependable, another invisible attribute. God declared that the days and seasons and years would continue as long as the earth exists (Genesis 8:22) and to this day the heavenly bodies remain a testimony to God's veracity. God keeps His promises; He is a God of truth. So far, I have listed ten of God's invisible attributes revealed by the visible sky: He is powerful; He is a God of order; He is eternal; He is intelligent; He is supernatural, He is the Creator; He is the Giver of life, He is dependable, He keeps His promises; and He is a God of truth.

What does the visibility of the heavenly bodies have to do with WHICH creation theory is true? I don't know if I have the answer to that one either, but I can still play the "What-If" Game. **What if God wanted to reveal His invisible attributes that were most important to fallen man?** God is powerful, God is intelligent, God is eternal, etc., and these things are seen in the sky regardless of which theory of creation is correct. But, if the Gap Theory is true, then the skies reveal something about God that the Day-Age and Young-Earth Theories don't reveal: **God is a God of Restoration.** If the Gap Theory is true, then when we look at the heavens, we see that God is a God of Judgment AND He is a God of Forgiveness. The earth was destroyed by His judgment, but He is a Merciful God and a Compassionate God. He is a God whose plans cannot be thwarted. His purpose for the earth will be accomplished. Sin can't defeat Him. Sin didn't take Him by surprise. Lucifer and his fallen angels didn't destroy God's plan for creation. These attributes are revealed by the heavenly bodies because God lifted His judgment and caused their light to shine into the darkness caused by sin. If you didn't know the earth was ruined and restored, then you wouldn't see these invisible attributes of God in the visible heavens. You might see His power, but you wouldn't see His forgiveness. You might see His intelligence, but you wouldn't see His redemption. None of His merciful and compassionate attributes are visible in the sky if the earth had only one beginning. But, if there were two beginnings, then the heavens declare these additional attributes of God. They declare that the God who judges is also the God who redeems. Yes, this is a "What-If" proposal, but how does it compare to the "What-If" proposals of the other theories? What do the heavens declare if the Young-Earthers are correct? They declare that we can't trust what we see.

He's not trustworthy. Stars that were never there and never exploded were made to look as if they were there and had exploded. Because His handiwork is deceptive, the stars reveal that God is deceptive. What do the heavens reveal if the Day-Agers are correct? They reveal that God's Word and His Works don't agree. In His Creation, He created the sun BEFORE the plants and trees, but in His Revelation He got it mixed up and told us He created the sun AFTER the plants and trees. The sky of the Day-Age Theory reveals that God is a God of confusion. Unlike those theories, the Gap Theory has "What-Ifs" that give God even more glory.

I believe Adam knew he was taken from the earth. I believe Adam knew the earth had been judged and restored. Adam looked at his own fallen nature and at the judgment God had pronounced upon him. He looked at the earth, then he looked at the sky, and he understood that God was a God of Restoration, Redemption, Compassion, Mercy, and Forgiveness. I believe that Adam, the man made from the dust of the earth, knew he would ultimately share in the fate of the earth. I believe Adam knew that someday he would be restored. I think God purposefully planned the Ruin and Restoration of the earth so that we would be able to look at the heavens and see Him as Our Savior, Redeemer, and Restorer from sin and death. These are the invisible attributes of God that are revealed in the visible heavens ONLY IF THE GAP THEORY IS TRUE. This is the special mission, the assignment, the *NATHAN*, He bestowed on the sun and the moon and the stars when He "set" them in the heavens. They are ambassadors telling us that God restored light and life to earth. He could have left earth dark and dead forever, but He didn't. In His love and mercy and forgiveness, He brought the earth out of the bondage of sin and death and into the light of new life. **If the other creation theories are true, then you can't see God's attributes of redemption and forgiveness in the sky.** Look at the stars. If the Gap Theory is true, then the only reason you see them is because God restored their light. Look at the moon. If the Gap Theory is correct, then you can see it only because God redeemed our planet from the sin and death that Lucifer brought upon it. Look at the sun. (But don't look too long.) If the Gap Theory is true, then you can understand that it was God's Grace that caused its light to shine where the darkness of sin had once ruled. If the Gap Theory is not true, you can't see God's forgiveness in the visible sky. You can't see any of His redemptive attributes. The Young-Earth Theory doesn't reveal that God is a God of forgiveness. The Day-Age Theory doesn't reveal that God is a God of restoration. The Gap Theory gives testimony to the most important thing we sinners need to know about God. God is a God of mercy, compassion, grace, forgiveness, redemption, and most of all restoration. We know He is the God who restores because we see in Genesis 1:3-31 that He restored creation. God is the God who redeems and restores that which was lost. God can restore life to what once was dead. Rather than being angry at the idea that God created a very good world on "top of a heap of bones," we ought to rejoice that He is the kind of God who would do that very thing. God is the God who restores! God is the God who forgives! If the Gap Theory is true, sinners can have hope in the God of Genesis 1:3-31. **Instead of creating conflicts with the Gospel of Jesus Christ, the Gap Theory demonstrates the Gospel of Jesus Christ.**

I believe God is smart enough and powerful enough and sovereign enough to make His Works and His Words agree. God's **Word** reveals His redemptive attributes. Why shouldn't His **Work** reveal His redemptive attributes? Why would God reveal His power, sovereignty, and intelligence in the heavenly bodies but not His mercy, grace, and forgiveness? If the Gap Theory is not true, then God doesn't reveal these attributes in the heavens. If the Gap Theory is not true, then God's Work doesn't reveal what His Word reveals. His Works don't match His Words. Yes, we need to see all of God's attributes, but the attributes we sinners need to see most are His attributes that cause us to see Him as our Savior. Even the fallen angels know that God is a God of omnipotence, omniscience, and omnipresence, but they don't see Him as a God of forgiveness and mercy and grace. **All people in all cultures and in all times need to see these attributes. So, God reveals them in the heavens for all to see.** There is no place on earth where people cannot understand that God is a God of restoration. All they have to do is look at the sky and realize that the celestial bodies are visible only because God restored them. Unfortunately, the Gap Theory has been rejected and forgotten for so long that the world has been blinded to everything the heavens fully reveal about God. I believe this is the work of Satan. Above all, Satan does not want us to see God as a forgiving God. When other creationists speak out against the Restoration Theory, I believe they are unintentionally helping Satan blind the world to the fact that God is our Restorer and Redeemer. They diminish the glory of God. Which creation theory gives God the greatest glory and honor and praise and majesty? Cosmogony recapitulates soteriology. That's the theory I believe!

Chapter Nine:

The Original Earth

—w—

Isaiah 45:18 provides us with the knowledge that God did not create earth a desolate waste. This should be enough to convince us that God did not create earth a desolate waste, but some theologians still disagree. So, let's look at another account of the early earth. God has provided us with another glimpse of earth prior to Genesis 1:3. This glimpse is found in Jeremiah 4:23-26. The context of Jeremiah 4 is a warning to Israel. God is warning the Jews about their idolatry and their sins. They have rebelled against God. From verse one to verse twenty-two, God reminds them of their wickedness and calls for their repentance. If they don't repent, they risk God's judgment and wrath. Suddenly, starting at verse twenty-three, Jeremiah begins a series of four "I beheld..." statements followed by physical descriptions of the earth. The first of these statements reads as follows according to the various translators:

Jer. 4:23 "I beheld the earth, and, lo, *it was* without form, and void; and the heavens, and they *had* no light." (KJV)

Jer. 4:23 "I beheld the earth, and indeed *it was* without form, and void; And the heavens, they *had* no light." (NKJV)

Jer. 4:23 "I looked at the earth, and it was formless and empty; and at the heavens, and their light was gone." (NIV)

Jer. 4:23 "I looked on the earth, and lo, it was waste and void; and to the heavens, and they had no light." (RSV)

Jer. 4:23 "I looked on the earth, and lo, it was waste and void; and to the heavens, and they had no light." (NRSV)

Jer. 4:23 "I looked at the earth. It was empty and had no shape! I looked at the sky. And its light was gone." (NCV)

Jer. 4:23 "I beheld the earth, and see, it was formless and empty, and the heavens had no light." (NBV)

Jer. 4:23 "I looked to the earth—it was a formless waste; to the heavens, and their light had gone." (NJB)

Jer. 4:23 "I looked at the earth, and lo! it was chaos; At the heavens, and their light was gone." (S&G)

Jer. 4:23 "I looked on the earth, and behold, it was formless and void; and to the heavens, and they had no light." (NASB)

Jer. 4:23 "I looked at the earth, and saw it was formless and empty. And the heavens, they had no light." (SCRIP)

Jeremiah describes what God revealed to him about the earth. It is *TOHUW WAW BOHUW*. Jeremiah describes the earth using the very words of Moses from Genesis 1:2. He is seeing earth at a time when it is without form, and void, and dark. This is important because **THERE IS ONLY ONE TIME IN HISTORY WHEN THE EARTH WAS *TOHUW WAW BOHUW* AND DARK.** It was *TOHUW WAW BOHUW*, a desolate waste, dark, and devoid of life only during the time before the six days of Genesis. Since the creation of Adam, the earth has never been a desolate waste, without light, and devoid of life. More importantly, it will never be in such a condition in the future. The heavens and the earth will pass away, as described in 2 Peter 3:10, but they will pass away only in the sense of being remade. They will not be unmade. The earth will be remade with a great, fervent heat, but it will not become devoid of life, and it will not be made into a desolate waste. Yes, at the Final Judgment, the earth will be darkened, but it will not be without form, and void, and it will not be dead. It will never again be *TOHUW WAW BOHUW*. Jeremiah is seeing earth during the time prior to Genesis 1:3, the only time in its history when it was *TOHUW WAW BOHUW*. Now, just so I'll know that you know what I know, I'm going to repeat this. The earth has never been *TOHUW WAW BOHUW* since Adam. The earth will never again be *TOHUW WAW BOHUW* according to the Bible. The earth was *TOHUW WAW BOHUW* only once. Therefore, if God reveals a glimpse of earth in its *TOHUW WAW BOHUW* condition, then it must be a glimpse of earth during the only time it was *TOHUW WAW BOHUW*. Jeremiah is looking at the past, not the future.

The question must be asked. What does the early earth in a state of darkness and desolation have to do with Israel being judged by God's wrath? Why does Jeremiah insert a description of the Pre-Adamic earth among the warnings to Israel? I purposely left off the rest of Jeremiah's description of the earth so you could first see how similar Jeremiah 4:23 is with Genesis 1:2. I did this because Jeremiah goes on to reveal things about the Pre-Adamic earth that Genesis doesn't reveal. Remember, it's okay for God to do this. Very few topics in the Bible are exhaustively covered in only one portion of Scripture. The creation account is no different. Jeremiah tells us something more about the early earth. It's important to understand, however, that Jeremiah doesn't contradict what Moses told us in Genesis. Jeremiah simply gives us additional information about the earth prior to the six days of restoration. God tells us something through

Jeremiah that shakes up the whole picture of earth's history. Jeremiah is about to say something you may have not known before. There was life on earth before the six days of Genesis. Jeremiah 4:23-26 is the key to understanding the relationship between Genesis 1:1 and Genesis 1:2. This key allows us to translate Genesis informationally, not just grammatically.

Jer. 4:23-26 "I beheld the earth, and, lo, *it was* without form, and void; and the heavens, and they *had* no light. *{24}* I beheld the mountains, and, lo, they trembled, and all the hills moved lightly. *{25}* I beheld, and, lo, *there was* no man, and all the birds of the heavens were fled. *{26}* I beheld, and, lo, the fruitful place *was* a wilderness, and all the cities thereof were broken down at the presence of the LORD, *and* by his fierce anger." (KJV)

Jer. 4:23-26 "I beheld the earth, and indeed *it was* without form, and void; And the heavens, they *had* no light. *{24}* I beheld the mountains, and indeed they trembled, And all the hills moved back and forth. *{25}* I beheld, and indeed *there was* no man, And all the birds of the heavens had fled. *{26}* I beheld, and indeed the fruitful land *was* a wilderness, And all its cities were broken down At the presence of the LORD, By His fierce anger." (NKJV)

Jer. 4:23-26 "I looked at the earth, and it was formless and empty; and at the heavens, and their light was gone. *{24}* I looked at the mountains, and they were quaking; all the hills were swaying. *{25}* I looked, and there were no people; every bird in the sky had flown away. *{26}* I looked, and the fruitful land was a desert; all its towns lay in ruins before the LORD, before his fierce anger." (NIV)

Jer. 4:23-26 "I looked on the earth, and lo, it was waste and void; and to the heavens, and they had no light. *{24}* I looked on the mountains, and lo, they were quaking, and all the hills moved to and fro. *{25}* I looked, and lo, there was no man, and all the birds of the air had fled. *{26}* I looked, and lo, the fruitful land was a desert, and all its cities were laid in ruins before the LORD, before his fierce anger." (RSV)

Jer. 4:23-26 "I looked on the earth, and lo, it was waste and void; and to the heavens, and they had no light. *{24}* I looked on the mountains, and lo, they were quaking, and all the hills moved to and fro. *{25}* I looked, and lo, there was no one at all, and all the birds of the air had fled. *{26}* I looked, and lo, the fruitful land was a desert, and all its cities were laid in ruins before the LORD, before his fierce anger." (NRSV)

Jer. 4:23-26 "I looked at the earth. It was empty and had no shape! I looked at the sky. And its light was gone. *{24}* I looked at the mountains, and they were shaking! All the hills were trembling. *{25}* I looked, and there were no people! Every bird in the sky had flown away. *{26}* I looked, and the good, rich land had become a desert! All its towns had been destroyed. The Lord and his great anger has caused this." (EB)

Jer. 4:23-26 "I beheld the earth, and see, it was formless and empty, and the heavens had no light. *{24}* I beheld the mountains, and look! They trembled, and all the hills were in commotion. *{25}* I looked, and see! There was no man, and all the birds of heaven had fled. *{26}* I looked, and behold! The garden land was a desert, and all the cities were broken down before the Lord in the presence of His fierce anger." (NBV)

Jer. 4:23-26 "I looked to the earth—it was a formless waste; to the heavens, and their light had gone. *{24}* I looked to the mountains—they were quaking and all the hills were rocking to and fro. *{25}* I looked—there was no one at all, the very birds of heaven had all fled. *{26}* I looked—the fruitful land was a desert, all its towns in ruins before Yahweh, before his burning anger." (NJB)

Jer. 4:23-26 "I looked at the earth, and lo! it was chaos; At the heavens, and their light was gone. *{24}* I looked at the mountains, and lo! they were quaking; and all the hills swayed to and fro. *{25}* I looked, and lo! there was no man, and all the birds of the air had flown. *{26}* I looked, and lo! the garden land was desert, and all its cities were ravaged before the Lord, before his glowing anger." (S&G)

Jer. 4:23-26 "I looked on the earth, and behold, it was formless and void; and to the heavens, and they had no light. *{24}* I looked on the mountains, and behold, they were quaking, and all the hills moved to and fro. *{25}* I looked, and behold, there was no man, and all the birds of the heavens had fled. *{26}* I looked, and behold, the fruitful land was a wilderness, and all its cities were pulled down before the Lord, before His fierce anger." (NASB)

What is God telling the Jews? He's warning them about His judgment and wrath by calling to mind a historical fact about the early earth that they understood. They knew God had judged the earth before Adam. God reminded them that in His judgment and fierce anger, He made the earth a desolate waste, devoid of life. He's warning them that unless they repent and forsake their evil ways, He will make them the same. If they do not cease their rebellion, He will make Israel desolate just like He made the early earth desolate, but unlike the early earth, He will not make a full end of them (vs. 27).

Jer. 4:27 KJV "For thus hath the LORD said, The whole land shall be desolate; yet will I not make a full end." (KJV)

Whoa! When did God have a great, fierce, burning, glowing anger at the earth? When did He make the earth *TOHUW WAW BOHUW*? When did He make a full end of the earth? When has there been no man and no light on earth? This can only be describing earth prior to Adam. Jeremiah is using *TOHUW WAW BOHUW*, the same words God used to describe the Pre-Adamic earth in Genesis 1:2. In both passages earth is desolate and lifeless. Both passages say the earth is dark and without light. Jeremiah must be looking at the Pre-Adamic world because from the time of Adam until all eternity, the earth will be populated by man. Earth will never be without man. Man became a permanent fixture on earth the day God created Adam from its dust. If earth ever becomes devoid of people, then God's promise to the Jews will have been broken.

Jer. 31:35-36 "This is what the LORD says, he who appoints the sun to shine by day, who decrees the moon and stars to shine by night, who stirs up the sea so that its waves roar— the LORD Almighty is his name: {36} 'Only if these decrees vanish from my sight,' declares the LORD, 'will the descendants of Israel ever cease to be a nation before me.'" (NIV)

God promised the Jews that they will never cease as long as the earth exists. Since God's promises can't be broken, earth will never be without the Jews. At no time in the future will earth be without Jews. This means earth will never be without man. Jeremiah can only be describing the Pre-Adamic earth because that's the only time there was no man. Earth was *TOHUW WAW BOHUW* before Adam, but will never be that way again. At the beginning of Genesis 1:2 is the only time earth was a desolate waste with no man and no light.

What else did Jeremiah see in this earth-before-Adam? He saw that the gardens or fruitful places had been made into desolate wilderness. "Fruitful places," "garden land,"

and "rich land" describe such things as gardens, fields of grain, orchards, and vineyards. Jeremiah reminded the Jews that God in His anger turned these fruitful places into wasteland. But, that means there was life on earth before it became *TOHUW WAW BOHUW*. It was not *TOHUW WAW BOHUW* because God hadn't yet created fruitful places; it became *TOHUW WAW BOHUW* because God destroyed the fruitful places that already existed. The creation of plants on day three of Creation (Genesis 1:11-13) was a restoration of plants. Jeremiah also saw that birds had lived on the earth before it became *TOHUW WAW BOHUW*. The creation of birds on day five of Creation (Genesis 1:20-23) was a restoration of birds. I can only wonder about what kind of plants and birds these were, but what really grabs my attention is the mention of cities. Jeremiah looked and saw that all of earth's cities had been pulled down and destroyed by God's wrath.

When have all of earth's cities been destroyed? Was this the Great Flood of Noah's time? It's very tempting to think so, but there are two major differences between the earth that Jeremiah sees and the earth at the time of the Great Flood. The earth wasn't dark during The Flood. Earth wasn't without people during The Flood. There were eight souls who did not perish. It would be wrong for Jeremiah to say there was no man; there were four men and four women on the earth. Jeremiah is not looking at earth at the time of The Flood. Jeremiah is not warning Israel that God will destroy them with a great flood. God had promised never to do that again. Instead, Jeremiah is warning them with a description of the earth made dead and desolate by the wrath of God. Is this a description of earth's future? No, if this applies to the future, then all the cities, including Jerusalem, will be destroyed, and this will never happen according to the Bible. There will never be a time in earth's future when all of its cities will be destroyed. The destroyed cities that Jeremiah sees must have been Pre-Adamic.

233

To *HAYAH* or not to *HAYAH*, that is the Question

There is one aspect of Jeremiah's vision of the early earth that may seem troublesome for the Gap Theory. Jeremiah doesn't use *HAYAH* ("was" is in italics) when he describes what he SEES, whereas Moses uses *HAYAH* when he describes what God is DOING.

Jer. 4:23-24 "I beheld the earth, and, lo, *it was* without form, and void; and the heavens, and they *had* no light. *{24}* I beheld the mountains, and, lo, they trembled, and all the hills moved lightly. (KJV)

Gen. 1:2 "And the earth **was** without form, and void; and darkness *was* upon the face of the deep. And the Spirit of God moved upon the face of the waters." (KJV)

Some opponents of the Ruin-Restoration Theory have tried to use this difference as a way of proving that Genesis 1:2 describes a static condition. Arthur Custance does an excellent job of refuting their argument. He shows how the ancient Hebrew people thought when *HAYAH* was or wasn't used. Custance reminds us that Jeremiah is seeing a point in time. He is not seeing a complete replay of the creation. When used to describe what he SAW, Jeremiah would not have used *HAYAH* because he is describing a static point. He is merely describing what God showed him. On the other hand, Moses is not just describing what God is SEEING; he is describing what God is DOING. Genesis is not merely a static description of what you would see if you were present at the creation. Genesis is a dynamic description of God's actions. This is obvious since Genesis says, "God created," "God said," "God moved," "God called," "God made," "God formed," etc. The emphasis is on God's actions. Jeremiah's account shows how Genesis 1:2 would have been written if

God had intended it to be a static description of what WAS instead of a dynamic description of what WAS COMING TO BE. Jeremiah's account clarifies how the ancient Hebrews used *HAYAH* when they thought of "being" in a dynamic sense, but didn't use it when they thought of "being" in a static sense. Custance explains this.

"Indeed, this is precisely what Jer. 4.23 does. Jeremiah's vision was a vision of a moment. He saw the earth as a Chaos. More than this, he saw a Cosmos as a Chaos, for he actually says that the evidence of civilization lay in ruins.... , men and cities had been overwhelmed. He was not concerned in reverting to the past in order to say that this scene of devastation had come about over a period of time by such-and-such a process. He merely says that when he saw it, it presented to his mind's eye a scene of devastation. It is almost as though the Author of Scripture had given us this passage in order to assist us in our understanding of Gen. 1.2 which so nearly parallels it while at the same time differing from it in such an important detail – the introduction of היה."[43]

Once again I want to make this clear. Jeremiah sees the earth in its history before there was man. He sees the earth as a desolate waste, devoid of life and light. Furthermore, this desolate condition was not the result of God not having yet created life and light. Instead, it was the result of God's judgment. Jeremiah is telling the Jews about God's judgment. He's warning them about God's judgment. That's the theme of this passage. If the *TOHUW WAW BOHUW* condition of the earth was not the result of God's Judgment, then it makes no sense for Jeremiah to mention it. Jeremiah connects it to God's judgment because there had been life, but it was destroyed. There had been light, but God took away that light. God judged the earth. Earth's life was destroyed and its cities were thrown down. These passages do not describe the earth at any time in its history since Adam, or anytime in its future. Neither do they describe Israel (The Land) at any time in history. Jeremiah must be seeing earth's past, prior to the restoration. The best interpretation of Genesis One is to let Scripture interpret Scripture and realize that the six days of creation were six literal days in which God restored the earth and the life on it.

Is Jeremiah 4:23-26 Literal or Figurative?

There are those who insist that Jeremiah is being figurative, i.e. the "world system" was useless, vain, corrupt, etc., and was in "spiritual darkness." Once again I disagree. There was no "world system." There could be no "world system" without men, and there was "no man." Men weren't created yet. The "world" is not symbolic of anything. Jeremiah is describing the literal world. Jeremiah is being literal when he tells us about the literal sin and the literal rebellion of the literal Jews. He's being literal when he tells us that the literal God is literally angry with the literal Jews and that He is literally going to bring literal judgment and literal destruction down of them if they don't literally repent and literally turn back to Him. I have literally read commentaries stating that since Jeremiah is writing in poetry here, this portion of Scripture can't be taken as being literal; it must be metaphorical or allegorical. What a literally ridiculous comment. Are we to believe that in all of human literature there has never been a poem that expressed literal ideas or thoughts?

Some will say that we can't take Jeremiah's description of earth as literal. Some will say that he was using spiritual symbolism as a way to convince the Jews to repent of their wicked ways and return to the Lord. Some will say that references to worldwide darkness, worldwide destruction of plant life, and worldwide destruction of cities are merely poetic tools used to express God's wrath. Again, as in Isaiah, there is nothing that indicates these things are anything but literal. The only reason for not accepting them as literal is because if they are literal, then certain pet interpretations of Genesis have to be discarded. I cannot accept these God-inspired words of Jeremiah to be figurative symbols of a nonexistent reality. Look at all the other warnings God gave Israel. God ALWAYS WARNED THEM WITH LITERAL CONDITIONS. When God warned Moses about Israel wandering around in the wilderness for forty years, that is exactly what happened. When God warned Israel that idolatry would lead to their destruction by foreign nations, they were literally destroyed by foreign nations. When God warned the Jews about being taken into captivity to Babylon for seventy years, they were taken into captivity to Babylon for seventy years. When Jesus wept over Jerusalem and warned his people of impending judgment and destruction, that is exactly what Rome did. **God's warnings to Israel have never described non-literal conditions.** When Jeremiah warned Israel of how utterly destroyed Israel would be made by God's wrath, he refers to a time in earth's history when God made earth *TOHUW WAW BOHUW*. I can only interpret this to mean that there was a literal time in earth's history when God made it literally *TOHUW WAW BOHUW*. To interpret *TOHUW WAW BOHUW* in a figurative sense here would be ridiculous. God didn't refer to figurative symbols when He warned Israel of judgment. Instead, He warned them with true, physical destruction, and in this instance He referred to the Pre-Adamic earth as His example. Divine judgment fell on the Pre-Adamic earth and God made it literally *TOHUW WAW BOHUW*.

The original earth was not a desolation. Isaiah said the earth wasn't created that way. Jeremiah said it was made that way after God, in His wrath, destroyed the life and the cities already there. Genesis 1:1 and 1:2 cannot be translated in such a way to imply that earth was originally created without form, and void. Even if the grammar allows it, such a translation must be laid aside in favor of an equally valid translation that doesn't introduce a contradiction into the Bible. If God originally created earth *TOHUW*, then Isaiah lied to us. If there were no plants and birds and cities in the Pre-*TOHUW WAW BOHUW*-earth, then Jeremiah lied to us. If these men lied, then God lied. God may have withheld information from us. I have no problem with that, but I can't allow God's character to be smeared by those who make Him out to be a liar simply because they wish to cling to their own preconceived ideas of what He could or couldn't do with His earth. God is not a liar; men are!

Does God Speak Hebrew?

Can we be sure that *TOHUW* and *BOHUW* imply Divine judgment in Genesis 1:2? According to Jeremiah 4:18, these words refer to Divine judgment. Jeremiah warns Israel that they will be made *TOHUW WAW BOHUW* if they don't repent. Are *TOHUW* and *BOHUW* used elsewhere? Yes, they are also found in Isaiah 34:11.

Isa. 34:8-11 "For the LORD has a day of vengeance, a year of retribution, to uphold Zion's cause. *{9}* Edom's streams will be turned into pitch, her dust into burning sulfur; her land will become blazing pitch! *{10}* It will not be quenched night and day; its smoke will rise forever. From generation to generation it will lie desolate; no one will ever pass through it again. *{11}* The desert owl and screech owl will possess it; the great owl and the raven will nest there. God will stretch out over Edom the measuring line of chaos (*TOHUW*) and the plumb line of desolation (*BOHUW*)." (NIV)

In Isaiah 34, God speaks of His judgment of all the nations and especially of Edom. Isaiah 34:11 tells us what

God will do to Edom when He judges it. God uses the words *TOHUW* (chaos) and *BOHUW* (desolation). Theologians and Hebrew scholars have been arguing for years over what these words mean in Genesis 1:2. How many times have you heard biblical scholars say that we need to let Scripture interpret Scripture? You've probably been bombarded with that phrase a gazillion times. Everybody claims they want to let Scripture interpret Scripture, but what they really mean is to let THEIR interpretation of Scripture interpret Scripture. What they're really saying is, "If your interpretation of Scripture looks like it's going to disprove my interpretation of Scripture, then you must be taking it out of context."

When we don't know the meaning of a word or phrase in the Bible, the first and best way of finding out its meaning is to go elsewhere in the Bible to see what it means. If we do that with *TOHUW WAW BOHUW*, then we discover that it means Divine judgment. It means Divine judgment in Jeremiah! It means Divine judgment in Isaiah! What would make me think that *TOHUW WAW BOHUW* has a different meaning in Genesis? What rule of biblical interpretation allows us to ignore the meanings of words and phrases found elsewhere in the Bible, just so we can interpret those words and phrases the way we want? Dear friends, a lot of people hate the Gap Theory. A lot of scholars claim that interpreting Genesis 1:2 as the result of Divine judgment is a stretched interpretation. They say it is weak and without merit. They say there is no biblical basis for believing it. I would beg to disagree. As far as I'm concerned, the burden of proof ought to be to show that Genesis 1:2 doesn't refer to Divine judgment. We have two passages of Scripture that use these words to describe Divine judgment. Jeremiah 4:18 and Isaiah 34:11 are the only verses besides Genesis 1:2 that use *TOHUW* and *BOHUW* together. In both those verses *TOHUW* and *BOHUW* refer to God's judgment. We ought to let Scripture really interpret Scripture. We ought to let words mean what they mean! We ought to let God say what He wants to say. God knew what He was going to say in the Book of Isaiah. He knew what the words would convey to Jeremiah. He knew all their meanings, shades of meanings, connotations, denotations, implications, and whatever. Hey, God speaks Hebrew! God chose two words to describe the Pre-Adamic earth and He picked the same two words that describe conditions of Divine judgment. Did God blunder in the selection of these two words? Didn't He know what He would imply by using them? I think God knew which Hebrew words to use to express what He wanted us to know. God seems to want us to know that Genesis 1:2 describes a Divine judgment.

Whose Cities Were These?

I know there was plant life, bird life, and I'll even speculate about animal life on the Pre-Adamic earth, but whose cities were these? Why did God destroy them in His wrath? Why is God reminding the Jews of what He did to the Pre-Adamic earth? He's warning them that they face the same destructive wrath. God is telling them that just as He destroyed the Pre-Adamic cities, He will destroy them, with the exception that He won't make a full end to them. Who built these cities that were destroyed before Adam was created? Two sub-theories have emerged. The first is that "men," some type of human or hominid creatures, existed before Adam. The other is that angels dwelt on the Pre-Adamic earth and built cities. Whether it was "men" or angels (or both) is unimportant as far as the Gap Theory itself is concerned. What matters is that THERE WAS INTELLIGENT LIFE ON EARTH BEFORE THE SIX DAYS OF GENESIS. There was a Pre-Adamic civilization. This means that Genesis 1:3-31 was not part of the original creation of the earth. Earth had an earlier beginning. I strongly believe that the six days of Genesis

1:3-31 were six twenty-four hour days. I am convinced that all life now on earth originated from those living things God restored during those six days. However, I am equally convinced there was life on earth prior to those six days, and that Pre-Adamic life was destroyed by God's judgment. I am also convinced that no living thing, including those things alive before Adam, got here by any process of evolution.

A Look at Satan's World

I am of the opinion that angels lived on earth and built cities prior to earth's destruction. I believe it was Lucifer and the fallen angels who caused earth to be made desolate and the universe to groan under the power of sin. I believe this because the Bible tells us about a judgment prior to Adam, but it says nothing about humans. I know this doesn't prove there weren't humans prior to Adam, but if Pre-Adamic humans existed, the Bible is silent about them. (Of course, God has every right to withhold information from us if He wishes.) On the other hand, the Bible tells us about the sin and rebellion of Lucifer. The Bible tells us that one-third of the angelic hosts were pulled into that rebellion. God doesn't want us to be ignorant about those rebellious angels. That's why I think they were the ones who brought about the judgment mentioned in Jeremiah 4:23-26. If they weren't, then God gives us no clues for understanding Jeremiah's revelation. If the Bible tells us about sin and rebellion prior to the six days of Genesis, then it must be referring to the only creatures mentioned in the Bible who sinned and rebelled prior to those six days. It must be referring to Lucifer and his fallen angels.

Now, where do I get the notion that Lucifer and the angels sinned and rebelled before the six days? The Bible says, "God saw every thing He had made, and behold, it was very good. And the morning and the evening were the sixth day."

If Lucifer had already sinned and rebelled before the sixth day, how could God say that everything He made was very good? Young-Earth creationists insist that the Gap Theory creates a contradiction here. Let's examine their argument.

The simplest interpretation of Genesis One, they say, is that everything, including angels, was created during those six days. The easiest interpretation is that Genesis 1:1 is the first creative step in a series of creative steps taking place over a six day period in which absolutely everything was created. This would be the most logical and straightforward interpretation of the Bible if there was only one beginning. This would be the case if the original creation described in Genesis 1:1 was the same creation as the creation described in Genesis 1:3-31. This would be the only explanation if there was only one creation. If the Bible revealed only one beginning, then I'd have to agree that Lucifer and the rest of the angels were created during that beginning. If "made" (*ASAH*) was always equivalent to "create" (*BARA*), then we could conclude that since everything God had MADE was very good, then everything God had CREATED was also very good. But if the six days of creation (restoration) were not the same as the original creation, then "very good" would be referring to the things God made during the six days. It wouldn't necessarily refer to everything He had previously created in Genesis 1:1. If the earth was restored by God during those six days, then "every thing He had made" refers to His works of restoration. If there were two different beginnings, then Genesis 1:31 refers to the second beginning, the restoration, but not to the first beginning, the original creation.

Go back and carefully read Genesis One and look for a pattern. AS HE CREATED, God repeatedly pronounced the things He created during the six days as, "good." He spoke light into existence and said light was, "good." He separated the water from the land so that dry land appeared, and He said that was, "good." He commanded plants to grow and they grew; He said that was, "good." When He set the sun and moon and stars in place, He said that was, "good." Sea creatures came into being by His command and He said that was, "good." The same was true for land creatures; He said that was, "good." Finally, after the creation of man and woman, God looked at what He had done and said it was, "very good." Do you see the pattern? Genesis 1:1-2 does not fit that pattern. God never said, "and it was good," after mentioning the creation of the formless and void earth. Instead, the words He used bring to mind the exact opposite condition. The earth was desolate and dead. There was no light, and the Spirit of God was moving or brooding over the deep. Even without knowing the exact meanings of *TOHUW*, *BOHUW*, and *CHOSEK*, I think we'd get the idea that the condition of the earth WASN'T GOOD. Genesis 1:1-2 do not fit into the six days. That's the whole point of revealing earth's dead condition. He wants us know that the earth needed to be reborn and that the Holy Spirit was the One who provided that rebirth. The earth had to be born again. It needed a new beginning that only the Holy Spirit could provide, and that's exactly what He did. The earth was reborn; it had a new beginning. God's proclamation of "very good" was referring to the earth and the life on it. It wasn't referring to the angels. The restoration of the physical realm did not include a restoration of the angelic realm. God was preparing a place for man.

Chapter Ten:

Earth's Two Beginnings

—⟶⟵—

Can I prove the earth had two beginnings? I think I can. Let's approach this problem first from the assumption that there was only one beginning. I'll assume that the six day period of creation was the one and only beginning, the one and only creation of everything, including the angels. I will assume there was no other beginning. In other words, "In the beginning God created the heavens and the earth," is the same thing as the six days of Genesis. I now have to check this assumption with biblical facts to test its validity. If the Bible gives no clues of two different beginnings, then this assumption is probably true. If the Bible gives clues that there were two separate and distinct beginnings, then this assumption is most definitely false. As it turns out, God has given us clues of two different beginnings. My first clue is found in Job 38. I've already mentioned part of what God told Job, but now I want to show you some more.

Job 38:4-7 "Where wast thou when I laid the foundations of the earth? declare, if thou hast understanding. {5} Who hath laid the measures thereof, if thou knowest? or who hath stretched the line upon it? {6} Whereupon are the foundations thereof fastened? or who laid the corner stone thereof; {7} When the morning stars sang together, and all the sons of God shouted for joy?" (KJV)

"The morning stars" and "the sons of God" are terms used in the Old Testament for angelic beings. God tells Job that the angels were eyewitnesses to the creation of the earth. They sang when God laid its foundation. This means they came before the earth was finished, before the six days of Genesis. If "laying the foundation of the earth," "measuring it," "stretching the line upon it," "fastening the foundations," and "laying the corner stone" all refer to the six days, the one and only beginning of the earth, then Lucifer and the angels were in existence prior to those events. They couldn't have witnessed any of those things if they weren't already there. They had to be created before the earth. If the earth was created during those six days, then the angels obviously had their beginning before the six days. Now, let me add a side note here that helps support the Restoration Theory. I can't imagine the angels, Lucifer included, shouting for joy just because a dead, dark, cold, lifeless, desolate, water-submerged speck of dust was created. I could see getting excited about the creation of the sun, the stars, and the quasars. They are spectacular in their power and beauty, but the creation of a desolate waste planet is hardly worth singing about. Because the creation of the earth caused the angels to sing and shout with joy, it must have been created in some condition other than *TOHUW WAW BOHUW*. The angels' joy over the creation of the earth confirms Isaiah's statement that the earth was not created a desolate waste. We know then, that to be eyewitnesses of earth's creation, the angels were created before the earth, before the six days of Genesis. This passage also indicates that all the angels were together and joyous. Most theologians agree that the period described by, "When the morning stars sang together, and all the sons of God shouted for joy," indicates the time before Lucifer sinned. Therefore, Job teaches us two things. The angels had their own creation, their own separate and distinct beginning before the earth was created, and Lucifer didn't sin until after the earth was created.

Okay, one clue isn't very convincing, especially if it deals with the creation of angels. I may have proven there were two beginnings, but so far I've shown only that the first beginning was the creation of angels. Most creationists probably don't care if the angels were created before the six days of Genesis. Angels are a separate type of being and live in a separate type of existence. Let's look at the next clue.

Eze. 28:12-15 "Son of man, take up a lament concerning the king of Tyre and say to him: 'This is what the Sovereign LORD says: "'You were the model of perfection, full of wisdom and perfect in beauty. *{13}* You were in Eden, the garden of God; every precious stone adorned you: ruby, topaz and emerald, chrysolite, onyx and jasper, sapphire, turquoise and beryl. Your settings and mountings were made of pure gold; on the day you were created they were prepared *{14}* You were anointed as a guardian cherub, for so I ordained you. You were on the holy mount of God; you walked among the fiery stones. *{15}* You were blameless in your ways from the day you were created till wickedness was found in you.'" (NIV)

The title, "King of Tyre," is a mysterious title for Lucifer, and so we see that he wasn't a sinner when he was created. God didn't create him that way. He didn't begin that way. He existed in a perfect condition for an unspecified period of time before he sinned. In fact, he even walked around in Eden before his rebellion. (I believe this was a Pre-Adamic Eden that got destroyed and later restored like the other fruitful places.) This means that Lucifer didn't sin until after the earth was created. This agrees with Job. Now, if there was only one period of creation, one beginning, then Lucifer was still perfect when God finished creating the heavens and the earth. Lucifer did not fall during the time that God was creating. Lucifer did not sin during the time period the Bible refers to as the beginning. Fine, that's no problem. Young-Earth creationists, Day-Age creationists, and just about Every-Other Creationists agree. Almost everybody puts Lucifer's rebellion sometime after day six. This is because God declared all His works very good on day six, and everybody but Gap Theory creationists says this includes Lucifer and the angels. Only the Gap Theory creationists contend that Lucifer had already fallen before day six, and that the words "very good" described only the restorative works of the six days. Let's move to our third clue. This is something Jesus said about Lucifer.

John 8:44 "Ye are of *your* father the devil, and the lusts of your father ye will do. He was a murderer **from the beginning**, and abode not in the truth, because there is no truth in him. When he speaketh a lie, he speaketh of his own: for he is a liar, and the father of it." (KJV)

"He was a murderer from the beginning." What beginning? Lucifer's beginning? Jesus said Lucifer was a murderer from the beginning, but Ezekiel says Lucifer had no iniquity when he was created. Jesus can't be saying that Lucifer was a murderer from his own beginning unless the Bible contradicts itself. Lucifer did not begin as a fallen angel! Lucifer began perfect. Was Jesus saying that Lucifer was a murderer from the beginning of the universe? Well, if there was only one beginning, then yes! If the six days of Genesis was the time everything began, then Lucifer was a murderer from the time everything began. Yet we know this can't be true. Job and Ezekiel have just shown us that Lucifer didn't fall until after the creation of the earth. Lucifer WAS NOT a sinner from the beginning, but Jesus still said he WAS a sinner from the beginning. The assumption that there was only one beginning starts to break down. Could Jesus be referring to some other beginning? According to Jesus, Lucifer was a murderer from the beginning. The Apostle John understood His Master's teaching, and taught this as well.

1 John 3:8 "He who sins is of the devil, for the devil has sinned **from the beginning**. For this purpose the Son of God was manifested, that He might destroy the works of the devil." (NKJV)

It is very interesting that the Greek word used for "beginning" by both Jesus and John is *ARCHE*. This word is used many times in reference to the creation.

Matthew 19:4 "And he answered and said, 'Have ye not read, that He who created *them* from the **beginning** MADE THEM MALE AND FEMALE." (NIV)

John 1:1-2 "In the **beginning** was the Word, and the Word was with God, and the Word was God. {2} The same was in the **beginning** with God." (KJV)

Heb. 1:10 "And, Thou Lord, in the **beginning** hast laid the foundation of the earth; and the heavens are the works of thine hands." (KJV)

2 Pet. 3:4 "And saying, Where is the promise of his coming? For since the fathers fell asleep, all things continue as *they were* from the **beginning** of creation." (KJV)

Since Ezekiel reveals that Lucifer wasn't created a sinner, then Jesus and John can't be referring to Lucifer's creation. Since Job reveals that Lucifer hadn't yet sinned when the earth was created, then Jesus and John can't be referring to the creation of the earth either. Lucifer must have fallen into sin after the earth was created but before the "beginning" that Jesus and John mention. What beginning are Jesus and John talking about? Let's look at Lucifer a little closer. Ezekiel mentions one beginning, the creation of Lucifer, a point of time in which Lucifer was perfect. Jesus and John mention another beginning, a point of time in which Lucifer was already a sinner. These can't be referring to the same beginning. If they are, then one part of the Bible contradicts another part of the Bible. This would have Jesus and John say that Lucifer was a sinner from the time he was created, while Job and Ezekiel say that Lucifer wasn't a sinner until later. But where does that leave us? Jesus and John must not be referring to the same beginning that Job and Ezekiel mention. The beginning that Jesus and John mention had to be after that beginning, after Genesis 1:1. So what beginning is Jesus talking about that was after Genesis 1:1? Jesus must know there was more than one beginning.

Let me repeat this. I want you to see what I'm getting at, and I know it's a difficult concept. Jesus and John said Lucifer was a sinner from the beginning. Was it the beginning of Lucifer? No, Lucifer was perfect in all his ways from the day he was created. Were Jesus and John talking about the beginning of the heavens and the earth? No, Lucifer was one of the angels shouting and singing for joy when the earth was created. We also see him later walking around on earth while still perfect. Since iniquity wasn't found in Lucifer until after some time had passed, we know that time was already in existence. Time can't pass if time doesn't exist. In addition, time can't exist if space doesn't exist; they are a continuum. Since time and space were created together in the beginning, then the universe had already begun before Lucifer fell. He was a sinner from the beginning, but this beginning couldn't be the beginning of the universe. If Genesis 1:1-31 describes the one and only beginning of the universe, then there is a problem with somebody's biblical testimony. If Lucifer was a sinner from the beginning, then Ezekiel and Job were wrong. If Lucifer wasn't a sinner from the beginning, then Jesus and John were wrong. The Gap Theory prevents this contradiction. Job and Ezekiel were right when they told us that Lucifer wasn't a sinner from the (first) beginning. Jesus and John were right when they told us Lucifer was a sinner from the (second) beginning. The earth had two beginnings.

The End of the Beginning

If the Gap Theory is false, then we have a biblical contradiction. Jesus and John said something about Lucifer that contradicts what Job and Ezekiel said. If we look closer at what Jesus and John said, we will discover something else. Even if Job and Ezekiel had said nothing about Lucifer, there is yet another discrepancy. It is a related discrepancy, yet it is distinct. This second discrepancy is not so much about Lucifer's fall, but about the chronology of Lucifer's fall. How does the timing play out? Jesus and John said that Lucifer was a sinner from the beginning. Okay, when did Lucifer first sin? It had to be after the creation of the earth but before the time he deceived Eve. Let's see if we can do some more investigating.

If we assume the Gap Theory is wrong, then we must assume that Lucifer didn't fall until after the six days. Again, this is what other creationists claim. They say that God's declaration on day six that everything was "very good" included Lucifer and the angels. On the other hand, we Gap Theory creationists say that this declaration included only the things He restored during those six days. We say that the six day period of restoration was the second beginning. We say that Lucifer was a sinner from the second beginning, but not from the first. His rebellion occurred between the two beginnings. Now, if the Gap Theory is wrong, then Lucifer couldn't have sinned until after the six days were finished. In fact, it couldn't have been on the seventh day either. The text says that God was pleased with all His work and that He blessed the seventh day and declared it holy. I don't think God would declare the seventh day blessed and holy if it was the day that Lucifer rebelled and introduced sin into the universe. I don't think God would be pleased with all His work if it included the introduction of sin.

According to Gap Theory creationists, Lucifer became a sinner before day one. According to other creationists, Lucifer couldn't have become a sinner until day eight at the earliest. I hope you see the problem with their logic. If Lucifer was a sinner from the beginning, as Jesus and John say he was, then Jesus and John made a mistake. The beginning was already over by day eight.

Genesis 1:1 says "the beginning" was the time period in which God created the heavens and the earth.
Genesis 1:31 was the end of that period.
Genesis 2:1 says the creation of the heavens and the earth were finished on day six. Genesis 2:2 says all of God's work was completed by day seven.
Genesis 2:3 says He rested from creating and making on day seven.
Genesis 2:4 says that Genesis One is the account of how the heavens and earth WERE CREATED. It's in the past tense. The creation week was finished.

"The beginning" ended on day seven. ANY TIME AFTER DAY SEVEN WAS NOT PART OF THE BEGINNING. If "the beginning" was the time period in which God created everything, then the beginning was over. If Lucifer didn't sin until after the beginning was over, then Jesus and John couldn't have said that he was a sinner from the beginning. The chronology is all messed up. If the original creation of the universe is the same creation as the six days of Genesis, then there is no other beginning, and we have a contradiction in God's Word. This can't be possible. The original creation of the heavens and the earth (Genesis 1:1) and the six days of creation (Genesis 1:3-31) must be two separate beginnings occurring at two different times. Lucifer wasn't fallen at the first beginning, the original creation of the earth, but he had already fallen before the second beginning, the six days of Genesis. The Gap Theory explains this.

Lucifer's Rebellion

I hope you aren't totally confused by all this because we aren't done looking at Lucifer's fall. We need to look at this in more detail because I did something you may not have noticed. I put an unwritten assumption into my argument, and to be fair I need to defend that assumption. If the assumption is wrong, the argument fails. If you didn't catch my assumption, here it is: I assumed that the phrase "from the beginning" is a reference to the creation. It certainly seems to be, but that doesn't make it so. *ARCHE* can be used for the beginning of anything. Most of the time it is used in connection with the creation, but it is also used to describe the beginning of Christ's earthly ministry, the beginning of Christ's miracles at Cana, the beginning of a feast when a host serves the best wine, and the beginning of other events. Sometimes it is used of Christ Himself. He is "the beginning and the end." (It is also translated "principalities" but that's used in a different sense, so don't let that throw you off.) Jesus and John said that Lucifer was a sinner from the beginning. We need to know what beginning they were referring to. The only way we can know that is to discover when Lucifer rebelled against God. If Lucifer rebelled against God before Adam was created, then my argument stands. If Lucifer didn't rebel against God until after Adam was created, then my argument falls. Are there more clues?

The Bible presents several clues that the earth had two beginnings. Like a jigsaw puzzle we can fit these pieces together to get a picture of what happened in the beginnings. This next piece of the puzzle concerns the timing of Lucifer's fall. When did Lucifer first rebel against God? When was iniquity first found in him? When did he become Satan, the devil? Was it before Adam was created or after? Gap Theory creationists hold that Lucifer fell before God created Adam. Other creationists believe that Lucifer fell after Adam was created. They don't believe God would say that everything was very good if sin and rebellion had already been present. They believe there was no sin, no death, and no judgment until after the creation week. They believe this declaration of "very good" applied to all of creation, including the angels. All of creation did not "groan" at this point. At the end of the sixth day, Lucifer and all the angels were still perfect, still shouting for joy, and still worshipping God. So, when did Lucifer sin? In order to answer that question we need to ask another question. When did God judge him, strip him of his authority, and cast him out of Eden? Let's look again at his original commission from God.

Eze. 28:13-16 "You were in Eden, the garden of God; every precious stone adorned you: ruby, topaz and emerald, chrysolite, onyx and jasper, sapphire, turquoise and beryl. Your settings and mountings were made of gold; on the day you were created they were prepared. *{14}* **You were anointed** as a guardian cherub, for so **I ordained you**. You were on the holy mount of God; you walked among the fiery stones. *{15}* You were blameless in your ways from the day you were created till wickedness was found in you. *{16}* Through your widespread trade you were filled with violence, and you sinned. So I drove you in disgrace from the mount of God, and I expelled you, O guardian cherub, from among the fiery stones." (NIV)

The text goes on to say other things about God's judgment on Lucifer, the King of Tyre. It becomes difficult to fully understand what God is saying too. Some of this appears to have already happened; some of it appears to be future. God goes on to say how Lucifer became proud because of his beauty and corrupt because of his splendor. It also talks about God casting him to the earth and turning him to ashes by a consuming fire that comes out from within him. We know this hasn't happened yet. Some of the confusion is probably because God is speaking about two different entities, in two different realms, at two different times. Two different experiences, one in the natural realm and one in the spiritual realm, seem to be blended into one. In Ezekiel 28:12, the Lord is speaking about the "KING of Tyre," but back in Ezekiel 28:2 the Lord was speaking about "the PRINCE of Tyre." I'm not sure I understand this, but it seems as if Lucifer is the King of Tyre while some man is the Prince of Tyre over whom Lucifer had control. Anyway, even if I can't explain all this, we have enough clues to proceed.

Lucifer appears to have been put in charge of Eden. We don't know how long he exercised that authority, but he eventually sinned because of his pride. He was then judged by God and stripped of his title. An important point in my argument is the chronology. God didn't strip him of his authority until after he rebelled. **As long as Lucifer remained perfect in all his ways, there was no need (or basis) for God to judge him or to cast him out.** So, when did this happen? I believe the Bible teaches that Lucifer sinned, was judged, and was stripped of his position of authority before Adam was created. If this happened before day six, then it would mean that Satan rebelled in the Pre-Adamic world. This in turn would mean that "very good" applied only to the six days of restoration, not to the previous creation. If this is true, then it would mean there was a Pre-Adamic civilization before our present world. All I need to do now is to prove that Lucifer sinned before Adam was created. I think I can do that, but we need some more clues. Like good detectives, we need to examine all the clues before we make our final accusation. (I say, "Lucifer did it; in the garden; with the forbidden fruit.")

Satan: Prince of this World

What do we know about Lucifer? God reveals many things about him, but He has chosen not to reveal everything about him. That's fine; God doesn't have to reveal any more than He wishes. There are things about the devil that God doesn't tell us in the Bible. Naturally, this has led many to make speculations about him. That's fine too. I think it is okay to try to fill in the missing pieces of the puzzle as long as we fully understand that we are only making guesses. Making guesses about things that AREN'T in the Bible is allowable, but we have to remember to see how they fit with the things that ARE in the Bible. I could speculate that Lucifer has red skin, horns, a pointed tail, and he carries a pitchfork. That's a very common image of Satan, but it's not from the Bible. It's wild, human speculation. The more we rely on what the Bible truly reveals about Lucifer, the less wild our speculations about him will be. Gap Theory creationists make some speculations about Lucifer. We speculate that Lucifer sinned and was judged before Adam was created. Other creationists also make speculations about Satan. They speculate that Lucifer didn't sin until after day six, after Adam was created. Let's see whose speculations fit with what the Bible tells us.

Let's first look at some of the things we know about Lucifer. He was created by God; he is a created being. Lucifer is not eternal. Lucifer is not omnipotent. Lucifer is not omniscient. Lucifer is not omnipresent. Simply put, Lucifer is not God. Now, he wants to be like God. He wants glory like God, but he is not and never will be God.

Isa. 14:13-15 "For thou hast said in thine heart, I will ascend into heaven, I will exalt my throne above the stars of God: I will sit also upon the mount of the congregation, in the sides of the north: *{14}* I will ascend above the heights of the clouds; **I will be like the most High**. *{15}* Yet thou shalt be brought down to hell, to the sides of the pit." (KJV)

We know Lucifer rebelled against God and has been judged, but we also know the final judgment hasn't been fully implemented yet. Lucifer is still free to roam the earth. He is even allowed to enter into God's presence in heaven at times.

Job 1:6-7 "One day the angels came to present themselves before the LORD, and Satan also came with them. *{7}* The LORD said to Satan, 'Where have you come from?' Satan answered the LORD, 'From roaming through the earth and going back and forth in it.'" (NIV)

We know his freedom is temporary. Someday he will be bound for a thousand years. After that, he will be set free to bring about one last battle against God. He will then be defeated by Christ, and he and all this followers, both angelic and human, will be cast into the Lake of Fire. There, for all eternity, they will be in continual torment for their continual hatred of God.

Mat. 13:41-42 "The Son of man shall send forth his angels, and they shall gather out of his kingdom all things that offend, and them which do iniquity; *{42}* And shall cast them into a

furnace of fire: there shall be wailing and gnashing of teeth."
(KJV)

The problem we need to address now concerns Satan's
authority over the earth. The Bible says that Lucifer is the
Prince of this world. This puzzle-piece needs to be put into
place.

1 John 5:19 "We know that we are children of God, and that
the whole world is under the control of the evil one." (NIV)

2 Cor. 4:4 "In whom the god of this world hath blinded the
minds of them which believe not, lest the light of the glorious gospel of Christ, who is the image of God, should shine
unto them." (KJV)

John 12:31 "Now is the judgment of this world: now shall
the prince of this world be cast out." (KJV)

John 14:30 "I will not speak with you much longer, for the
prince of this world is coming. He has no hold on me," (NIV)

Eph. 2:2 "Wherein in time past ye walked according to the
course of this world, according to the prince of the power of
the air, the spirit that now worketh in the children of disobedience:" (KJV)

John 16:8-11 "And He, when He comes, will convict the
world concerning sin, and righteousness, and judgment; {9}
concerning sin, because they do not believe in Me; {10} and
concerning righteousness, because I go to the Father, and
you no longer behold Me; {11} and concerning judgment,
because the ruler of this world has been judged." (NASB)

I don't fully understand what this means or why God allows it, but it is a clear declaration of God's Truth. Satan has authority over the earth. This played a central part in Satan's temptation of Christ.

Mat. 4:8-10 "Again, the devil took him to a very high mountain and showed him all the kingdoms of the world and their splendor. {9} 'All this I will give you,' he said, 'if you will bow down and worship me.' {10} Jesus said to him, 'Away from me, Satan! For it is written: 'Worship the Lord your God, and serve him only.'"" (NIV)

What was Satan offering Christ? Satan offered Jesus, the Son of David and legal heir to David's throne, the opportunity to set up His earthly Kingdom without having to go to the cross. Jesus chose to reject his offer. Instead, He chose to establish His earthly kingdom at the proper time and in the proper way. This Kingdom is what we refer to as the Millennial Reign of Christ. This is what Christ will set up when Satan is bound for a thousand years. I don't fully understand what it will be like. I know it will be wonderful and that Christ Himself will be physically present as King, but it will be a physical kingdom. It will be an earthly kingdom. It is not the final state of all things. The final perfection of the universe will not happen until after the final judgment. Once Satan is released, there will be one final rebellion, one final battle, and one final judgment. This is followed by a New Heavens and a New Earth, a perfect, eternal Kingdom, which I understand even less, but eagerly await for all the more.

Satan offered Christ the chance to set up His earthly kingdom without opposition. Lucifer would freely surrender his authority over the earth. Satan would give up his earthly kingdom in exchange for one, itty-bitty, little thing: Jesus would have to be willing to bow down before him. All Jesus had to do was acknowledge that Lucifer was worthy of worship, (Like the Most High) and then Jesus could be King of the earth. He wouldn't have to go to the cross. Satan seemed fine with losing his earthly kingdom. Why? Why would he make this offer? It seems almost ridiculous that Lucifer would willingly give up his authority over earth in exchange for something that seems so simple, even pointless. It seems even more ridiculous that Jesus wouldn't be willing to accept that offer. There must be more at stake than meets the eye. Something very important must have been happening in this confrontation. There was something going on that transcended just the sins of men. This wasn't just about who had authority over the earth. There was something deeper. Satan knew something very, very critical. Satan knew he could get something in exchange. Satan knew that God couldn't execute final judgment on him if Christ accepted his offer. **How could the Father condemn someone whom the Son had acknowledged as being worthy of worship?** You see, the cross wasn't just the focal point of human history; it was the focal point of angelic history as well.

1 John 3:8 "He who sins is of the devil, for the devil has sinned from the beginning. For this purpose the Son of God was manifested, **that He might destroy the works of the devil.**" (NKJV)

Satan knew he could escape eternal damnation, and he was willing to give up his authority over the earth in exchange for that freedom. Yes, Jesus could get His kingdom, but it would be the kingdom that Lucifer offered, not the Kingdom that God had planned. The earthly kingdom Lucifer offered, although full of splendor and wonder, would never be what God intended. It would be inhabited by fallen men whose sins were never forgiven. This kingdom would be a glorious kingdom, but it would be a kingdom of the unrestored. Ironically, this is the kingdom His disciples were seeking.

They were looking for an earthly kingdom in which they (of course) would be given great positions of authority and power. That's why Jesus rebuked Peter and called him "Satan" when he refused to let Christ go to the cross.

Mark 8:33 "But when he had turned about and looked on his disciples, he rebuked Peter, saying, Get thee behind me, Satan: for thou savourest not the things that be of God, but the things that be of men." (KJV)

Who Had Authority Over the Earth When?

My next question is this: How could Lucifer have authority over the earth? How could Lucifer, a created being, offer Christ, the Creator, authority over His own earth? We discover that our puzzle pieces are made of smaller puzzle pieces. We have the piece of that puzzle that says Lucifer has authority over the earth. He claimed to have that authority and Jesus didn't dispute it. This seems odd considering that Jesus was very quick to put Pilate in his proper place when he claimed to have authority over Jesus.

John 19:10-11 "Pilate therefore said to Him, 'You do not speak to me? Do You not know that I have authority to release You, and I have authority to crucify You?' {11} Jesus answered, 'You would have no authority over Me, unless it had been given you from above...'" (NASB)

If Satan did not have the authority he claimed to have, I think Jesus would have immediately said so. Jesus' lack of rebuttal to Lucifer's claim, along with the other Bible verses we have seen, leaves us with no interpretation but that Lucifer has real authority over the earth. He is its prince. He is its ruler. So, let me ask, "When did he get that authority?" Let me rephrase that question into three questions. The first question is, "When did he FIRST get that authority?" Once we answer the first question, we will discover there is a second question. "When did he LOSE that authority?" Once we answer that, a third question pops up, "When did he REGAIN that authority?" Look back at our previous puzzle piece.

Eze. 28:13-14 "You were **in Eden**, the garden of God; every precious stone adorned you: ruby, topaz and emerald, chrysolite, onyx and jasper, sapphire, turquoise and beryl. Your settings and mountings were made of gold; on the day you were created they were prepared. *{14}* You were **anointed** as a guardian cherub, for so **I ordained you**. You were on the holy mount of God; you walked among the fiery stones." (NIV)

The answer to the first question is easy. God gave Lucifer his position of authority in Eden. This was before Lucifer's fall. The Bible doesn't explain all that what was involved with this position of authority, but this is the passage of Scripture that describes it. He got this authority before he fell. He got this authority directly from God; God ordained him. This makes sense. He had to get this authority from God. All authority ultimately comes from God. This is God's universe.

1 Chr. 29:11 "Yours, O LORD, is the greatness and the power and the glory and the majesty and the splendor, for everything in heaven and earth is yours. Yours, O LORD, is the kingdom; you are exalted as head over all." (NIV)

Just as God gave authority to Pilate, God gave Lucifer authority. It was a much bigger authority than Pilate's. Pilate had authority over a very small piece of real estate for a very short time. Lucifer was given authority over the whole earth for a much longer time. God gave Lucifer authority over the earth, and this authority was given when Lucifer was still perfect in all his ways. But how does this fit with God giving

Adam authority over the earth? We know that God gave Adam authority over the earth as well. Who had authority over the earth when? There seems to be a problem with the chronology. Since Lucifer got his authority before he fell, it was obviously given to him before Adam fell. Lucifer was given this authority while he was still perfect in all his ways. This means it was given to him before he led Adam and Eve into sin. (Gap Theory creationists say that Adam wasn't even created yet.) This ordination happened when the universe was still perfect. This authority was given before sin was introduced. Now the problem with the chronology arises. Creationists who reject the Gap Theory say that sin wasn't introduced until after the end of day six, after God had finished with the creation. Unfortunately for their theory, by day six God had already stripped Lucifer of his authority over the earth and had given it to Adam. On DAY SIX God gave authority over the earth to Adam not to Lucifer.

Gen. 1:27-28... 31 "So God created man in his *own* image, in the image of God created he him; male and female created he them. *{28}* And God blessed them, and God said unto them, Be fruitful, and multiply, and replenish the earth, and subdue it: and **have dominion** over the fish of the sea, and over the fowl of the air, and over every living thing that moveth upon the earth.... *{31}* And God saw every thing that he had made, and, behold, *it was* very good. And the evening and the morning were **the sixth day**." (KJV)

Psa. 8:4-8 "What is man, that thou art mindful of him? and the son of man, that thou visitest him? *{5}* For thou hast made him a little lower than the angels, and hast crowned him with glory and honour. *{6}* **Thou madest him to have dominion over the works of thy hands; thou hast put all** *things* **under his feet:** *{7}* All sheep and oxen, yea, and the beasts of the field; *{8}* The fowl of the air, and the fish of the

sea, *and whatsoever* passeth through the paths of the seas."
(KJV)

Adam was given authority over the earth on day six.
Adam was given authority over the earth on the very day he
was created. Who was the ruler over the earth on day six? It
was Adam, not Lucifer. Do you see the problem? If Lucifer
had been ruler over the earth and had not yet sinned, then
God stripped Lucifer of his authority **before** he rebelled.
According to other creationists, Lucifer had not yet sinned
by day six. Yet, here we see that God had already stripped
him of his authority by day six. So, we have to ask the second
question about Lucifer, "When did he LOSE his authority?"
The answer is that he lost it **after** he had sinned. If Lucifer
had not yet sinned, then the Bible is wrong to say that God
stripped Lucifer of his authority and cast him out of Eden as
the result of his sin.

Eze. 28:13-17 "You were in Eden, the garden of God; Every
precious stone was your covering: The ruby, the topaz, and
the diamond; The beryl, the onyx, and the jasper; The lapis
lazuli, the turquoise, and the emerald; And the gold, the workmanship of your settings and sockets, Was in you. On the day
that you were created They were prepared. *{14}* You were
the anointed cherub who covers, And I placed you *there*. You
were on the holy mountain of God; You walked in the midst
of the stones of fire. *{15}* You were blameless in your ways
From the day you were created, Until unrighteousness was
found in you. *{16}* By the abundance of your trade You were
internally filled with violence, And you sinned; **Therefore** I
have cast you as profane From the mountain of God. And I
have destroyed you, O covering cherub, From the midst of
the stones of fire. *{17}* Your heart was lifted up because of
your beauty; You corrupted your wisdom by reason of your

splendor. I cast you to the ground; I put you before kings, That they may see you." (NASB)

God says, "Therefore I have cast you…." The "therefore" refers back to the fact that Lucifer had sinned. This judgment was the RESULT of his sin. God didn't judge Lucifer until after iniquity was found in him. Lucifer had been walking around in Eden, and he had been given authority over the earth. By the time we get to day six, we see that Adam is walking around in Eden. Adam has authority over the earth. Lucifer had already been stripped of his authority. Lucifer had already introduced sin into the universe BEFORE day six. God's pronouncement on day six that everything was very good did not include Lucifer. If there was only one beginning and one Garden of Eden, the chronology doesn't fit. Things become even more complicated when you try to fit this into the chronology of the Garden of Eden. When was the Garden of Eden planted? If there was only one Garden of Eden, then it was planted on day six at the earliest. The Garden of Eden was planted **after** Adam was created.

Gen. 2:8 "And the LORD God planted a garden eastward in Eden; and there He put the man whom He **had** formed." (KJV)

The Bible says God created Adam in Genesis 2:7. The Bible says God planted the Garden of Eden in Genesis 2:8. Genesis 2:7 comes before Genesis 2:8. If God is giving us a chronological account of creation, then Adam came before the Garden. God had already formed man before He planted the Garden of Eden. So if God didn't plant the Garden of Eden until after Adam's creation, then Lucifer couldn't have been walking around in the Garden until after Adam was created. But, if God gave Adam authority over the earth on day six, how could He give Lucifer the same authority over the same earth on the same day? It would seem odd for God to give Lucifer authority over the earth if He had already given that authority to Adam. Either God was lying to Lucifer when He told him he had authority over the earth, or else God was lying to Adam when He told him he had authority over the earth. They both can't have authority over the other. They couldn't have been given their authorities at the same time. Lucifer had already been stripped of his authority before day six. The only explanation is that earth had two Gardens of Eden. The first Garden of Eden was a Pre-Adamic garden that Lucifer walked around in and was given rule over. The second Garden of Eden was a garden that was planted after Adam was created. The second Garden of Eden was given to Adam to walk around in and to rule over. We now have the answer to my second question, "When did Lucifer lose his authority?" He lost it before day six, before the day Adam was created. Lucifer sinned, was judged, and lost his authority over the earth in the Pre-Adamic world.

None of this fits together if there was only one creation, one beginning, and one Garden of Eden. Earth had to have two beginnings. The only way to fit all these puzzle pieces together is to understand that the Bible tells us that there was an original creation, an original Garden of Eden, a rebellion by Satan, a judgment and destruction of the earth, and then a second creation. The Garden of Eden that Adam lived in was the restoration of the Garden of Eden that Lucifer lived in. It had been destroyed by God's wrath due to Satan's rebellion, but it was restored along with all the other fruitful places during the six days of restoration. My assumption about the phrase "from the beginning" was true. Jesus made that statement about Lucifer in reference to the second creation, but not to the first creation. He was referring to the beginning that took place over the six days of Genesis. Lucifer was a murderer from this beginning because undoubtedly, his heart was filled with hatred of Adam from the time Adam was created. Lucifer wanted Adam dead. In the eyes of Jesus, hatred is equal to murder. In the eyes of Jesus, Lucifer was a murderer from the beginning. The Gap Theory explains all this.

Lucifer Caused the Original Earth to Become
TOHUW WAW BOHUW

There was an earth that Lucifer walked around on before he fell. It was not yet *TOHUW WAW BOHUW*. The earth was beautiful and Lucifer walked around in the pre-Adamic Garden of Eden. How long he remained unfallen and how long the original earth remained beautiful is unknown, but by the time we get to Genesis 1:2, we see that earth was no longer beautiful. Lucifer's sin falls in the gap between Genesis 1:1 and Genesis 1:3. The *TOHUW WAW BOHUW* condition of the earth falls in that gap as well. In other words, Lucifer sinned during the same Pre-Adamic period that earth received a judgment for sin. This is why I think the condition of the earth was the results of God's judgment on the sins of Lucifer and the fallen angels. I hope you're beginning to see how all this fits together. Is there anything else? Is there another clue that might tie Lucifer's fall to the destruction and judgment of the earth between Genesis 1:1 and Genesis 1:3? Look at Isaiah 14.

Isa. 14:12-17 "How art thou fallen from heaven, O Lucifer, son of the morning! *how* art thou cut down to the ground, which didst weaken the nations! *{13}* For thou hast said in thine heart, I will ascend into heaven, I will exalt my throne above the stars of God: I will sit also upon the mount of the congregation, in the sides of the north: *{14}* I will ascend above the heights of the clouds; I will be like the most High. *{15}* Yet thou shalt be brought down to hell, to the sides of the pit. *{16}* They that see thee shall narrowly look upon thee, and consider thee, *saying*, Is this the man that **made the earth to tremble**, that did shake kingdoms; *{17} That* **made**

the world as a wilderness, and **destroyed the cities** thereof; *that* opened not the house of his prisoners?" (KJV)

Now this is earth-shaking! Look what Isaiah says Lucifer caused the earth to suffer. Lucifer made the earth tremble. He made the world a wilderness. He destroyed its cities. Does this sound familiar? Look back at what Jeremiah saw in the Pre-Adamic earth.

Jer. 4:24 "I beheld the mountains, and, lo, they **trembled**, and all the hills moved lightly..." (KJV)

Jer. 4:26 "I beheld, and, lo, the fruitful place *was* a **wilderness**, and all the **cities thereof were broken down** at the presence of the LORD, *and* by his fierce anger." (KJV)

It seems too great a coincidence for Isaiah and Jeremiah to have such similar visions of the earth if each vision describes earth at different times. They both are looking at the Pre-Adamic earth. Both Isaiah and Jeremiah see the earth trembling. Both prophets see the world made into a wilderness. Both see earth's cities broken down and destroyed. Jeremiah tells us that God did this out of judgment for sin. Isaiah tells us that Lucifer caused this. When did Lucifer do this? When did Lucifer make the earth a wasteland or wilderness? When did he destroy all the cities? He hasn't done it since Adam was created, and he won't do it in the future if God's promises mean anything. The world is not going to be destroyed when Christ returns to set up His earthly kingdom. The earth won't be destroyed when Satan is bound for a thousand years. The earth won't be destroyed when Satan is released and engages in one last battle. Christ is the victor, not Lucifer. Christ's kingdom will remain.

Mat. 13:41-43 "The Son of man shall send forth his angels, and they shall gather **out of his kingdom** all things that offend, and them which do iniquity; *{42}* And shall cast them into a furnace of fire: there shall be wailing and gnashing of teeth. *{43}* Then shall the righteous shine forth as the sun **in the kingdom** of their Father. Who hath ears to hear, let him hear." (KJV)

Satan, all fallen angels, and all unsaved people will be removed from the Kingdom at that time. The Kingdom remains, but Satan is removed. At the end of time, it will not be the earth that is destroyed; it will be Satan. Satan will not be able to destroy the earth in the future. The only time he could have done this is in the past. THE ONLY TIME LUCIFER COULD HAVE DESTROYED ALL THE CITIES AND MADE THE WORLD A WILDERNESS WAS BEFORE ADAM WAS CREATED. Again, the Restoration Theory fits these biblical facts better than the other theories.

What Did Lucifer Do?

I think angels inhabited the earth long before man. I think it was covered with beautiful gardens and it teemed with life. When the earth was created, it was so wonderful they all rejoiced. Earth must have been very beautiful. In fact, I think it's possible the entire universe was much different from what it is now. At that time, the whole of creation did not groan and was not in travail. At that time, the heavens and earth had not yet been shaken by God's judgment.

Suddenly, sin was introduced! Lucifer rebelled against God and pulled a third of the angels into his rebellion. I think he tried to create an empire of his own where he could have mastery. I think he wanted servants and slaves to worship him. It was his sin, not Adam's, that caused the whole universe to groan and be shaken. Lucifer began destroying the earth, the solar system, and the universe by starting a civil war among the angels. Most angels remained faithful to God, but the process of corruption had begun. Lucifer began corrupting and desolating the earth while using his power to twist things to his own purposes. He slowly turned earth into a cesspool of sin. I don't know how long this took, but he and his fallen angels continued to defile creation. Over time, their acts resulted in the death and extinction of many species of plants and animals. God finally intervened, and in His anger, destroyed the kingdom of the angels. He put an end to their universe. God judged the earth and made it *TOHUW WAW BOHUW*. The earth became without form, and void. He even may have changed the angels so that they could no longer freely inhabit the physical cosmos like before. Or, He could have changed the physical cosmos itself. The universe now groaned, and the earth was dark, dead and desolate.

Even in his defeat, Lucifer gloated over what he had done. He felt he had ruined God's universe and thwarted God's sovereign plan. I'm sure such thoughts gave him some kind of sick pleasure. If he couldn't be the god of this world, he at least made it so that God couldn't be God of this world. Lucifer couldn't have his way, but at least he had the pleasure of humiliating God by corrupting His creation. Oh, how beautifully ugly was sin's stain now smeared across the face of God's no-longer-perfect universe. At least he did something outside of God's plan... or so he thought. Little did he realize that God knew all along that he was going to rebel and corrupt all of creation. God knew this would happen. He had a plan for earth. God was going to redeem the earth. The earth was going to be born again, and God was going to resurrect the life that inhabited it. This time, however, there was going to be something different. This time there was going to be a new creature, a creature that would be no brute beast. This one was going to be made in the image of God, and be given dominion over the earth. (Just as the Christian is made into the image of Christ and becomes a co-heir with Christ.) I'm sure this galled Lucifer. He desired to have mastery over the earth, but he couldn't have it. Now, this puny creature, this man made from dirt, was going to have that authority. This creature was to be made lower than the angels, but God was going to make him ruler over the earth. God was going to make man, not Lucifer, a king. God was also going to show Lucifer that these lowly creatures would do what he and his rebellious angels wouldn't do. They would worship God. They would choose to obey God and love God, rather than rebel against Him. Although in power and in knowledge, man was far inferior to the angels, man would recognize that God alone is worthy of worship. The Dirt-Man would prove that Lucifer was not justified in seeking worship for himself.

God began restoring the earth. How fitting it was that His first act of restoration was to send light into the world. Darkness has always been used as a symbol of sin, and the first thing God did was to separate the light from the darkness. Years later, God would send another light into the world: Jesus, the Light of the world. He would separate the light from the darkness as well. It was equally fitting that God's last act of restoration was the creation of a new creature with whom He could have intimate fellowship. In a similar fashion, Jesus' last act was to go to the cross so we could become new creatures and have intimate fellowship with Him.

When Did Satan Regain Authority Over the Earth?

Lucifer lost his dominion over the earth when he was judged for his sin. How then could he have that dominion when he tempted Christ in the wilderness? Did God give it back to him as a reward for causing Adam and Eve to sin? That wouldn't make sense. That would be like rewarding a car thief with the title to your car because he was able to trick you into leaving the doors unlocked and the key in the ignition. If Adam lost his dominion because of sin, surely the one who committed the greater sin would not be given that dominion by God. So how could he claim to have authority over the earth? The answer is that someone gave him authority over the earth. How do I know that someone gave Lucifer authority over the earth? By now I hope you guess that I will say, "The Bible tells me so."

Luke 4:5-8 "The devil led him up to a high place and showed him in an instant all the kingdoms of the world. {6} And he said to him, 'I will give you all their authority and splendor, for **it has been given to me**, and I can give it to anyone I want to. {7} So if you worship me, it will all be yours.'" (NIV)

Someone gave the devil authority over the earth after he had been stripped of it. Who could have given him that authority? Surely God wouldn't reward Lucifer for causing Adam to sin! God had stripped Lucifer of his authority because of his own sin. Leading Adam and Eve into sin would not be a cause for reward. God didn't give Lucifer his authority back to him. But if God didn't give him that authority, who did? Here is the answer the Gap Theory proposes: Lucifer regained that authority when **Adam gave it to him.** According to the Gap Theory, God restored earth and on the sixth day put it under Adam's dominion. Satan couldn't stand it. I'm sure he hated Adam from the instant God announced He was going to create man in His own image. Lucifer began thinking of ways to destroy this new creature. Lucifer was a murderer from the beginning. He began plotting ways to cause man's death. I'm sure he watched Adam and Eve very closely, ever looking for a way to ruin God's plan again. How delighted Satan must have been when he deceived Eve. How much more delighted he must have been when Adam knowingly rebelled against God. Adam's sin brought death upon all his descendants and better still, from Lucifer's point of view, Adam's sin caused him to lose dominion over the earth. Lucifer had been able to conquer man, and in so doing, seized control of the world over man. Satan gained authority over Adam when Adam willingly joined Satan's rebellion. Lucifer started the rebellion, one-third of the angels submitted themselves to his rebellion, and then Adam and Eve joined his rebellion when they chose to believe Lucifer rather than God. Adam knowingly sided with the devil when he ate the forbidden fruit. In doing so, Adam handed his authority over to Lucifer. **Adam gave his authority to Lucifer by surrendering himself to him.** By obeying Lucifer rather than God, Adam made himself a slave to Lucifer.

Rom. 6:16 " Don't you know that when you offer yourselves to someone to obey him as slaves, you are slaves to the one whom you obey—whether you are slaves to sin, which leads to death, or to obedience, which leads to righteousness?" (NIV)

John warned us about this very thing when he put the ultimate effect of sin in its proper relationship with the events of the creation. Once again, here are John's words.

1 John 3:8 **"He who sins is of the devil,** for the devil has sinned from the beginning. For this purpose the Son of God was manifested, that He might destroy the works of the devil." (NKJV)

Adam made Lucifer his master by making himself his slave. He gave up his authority over the earth. It wasn't that Lucifer tricked Adam into leaving the keys in the car. Lucifer tricked Adam into signing over the title. Satan was now the prince of the world, and he set out to enslave men by the power of sin and death. It pleased Satan that Adam's sin led to a change in the earth. Thorns and thistles would plague mankind, and beasts would pose a threat. Man would have to live by the sweat of his brow, and suffer disease, pain, and death. Best of all, this precious little pet of God's was now going to be judged and cast into the Lake of Fire. Again, even though Lucifer knew he couldn't defeat God, he got pleasure out of knowing he could ruin God's plans.

But God knew about this too. He knew Adam would sin. He didn't want Adam to sin, but He knew he would. So God expelled Adam from the Garden and let death reign over man, but God had a plan for man. God was going to do something even the angels couldn't comprehend. He was going to become one of these puny creatures, one of these Dirt-Men. How strange it must have seemed to the angels. He had never taken on the nature of angels. The Son was now going to take on the nature of humans and live their miserable life and die their miserable death. More than that, He was going to bear their sins and die in their place. He was going to let the wrath of the Father be poured out on Himself for all the sins they had committed. Jesus would let the Father's judgment fall on Him rather than on His beloved children. This seemed impossible to imagine. God would let Himself bear the judgment of man's sin? Not even the angels could grasp what God had in store for man. What Lucifer couldn't understand was that through the Son's death, God would utterly defeat sin and destroy Satan's enslavement of man forever.

Heb. 2:14-16 "Forasmuch then as the children are partakers of flesh and blood, he also himself likewise took part of the same; that through death he might destroy him that had the power of death, that is, the devil; {15} And deliver them who through fear of death were all their lifetime subject to bondage. *{16}* For verily he took not on *him the nature of* angels; but he took on *him* the seed of Abraham." (KJV)

Through all this, God would raise up creatures who would someday be more than mere servants of God, they would be the sons and daughters of God. They would be more than just subjects to the King; they would reign with the King. They would become kings in God's Kingdom after all. No angel ever held that honor. Though men were made lower than angels, they would be raised up to a greater glory if and only if they accepted what God the Son did for them on Calvary's Cross. Those who are saved (Those who have claimed the Blood of Christ as their only hope) will be appointed by God as judges over the fallen angels. Satan planted the seeds of his own destruction on the day he first held iniquity in his heart, and he inadvertently selected his judges when he led Adam and Eve into sin.

1 Cor. 6:3a "Know ye not that **WE** shall judge angels?" (KJV)

If the Restoration Theory is true, then the earth's history typifies the history of the man made from its soil. The earth was created. Sin destroyed its original condition. This was followed by judgment and death. Next, there was a rebirth. Although reborn, the earth was still subject to sin and decay. It was very good, but not yet perfect. The earth awaits a future perfection. This is what happened to earth, and this is what happened to man. Man was created, but man fell into sin and judgment; he experienced death. The rebirth of the Christian is like the restoration of the earth. The restoration of the Christian is also very good, but not yet perfect. We too are still subject to sin and decay and are awaiting a future perfection. Both the earth and saved men wait for the day when they will be made perfect. Someday there will be a new heavens and a new earth, and saved men will receive new and perfect bodies. Someday both we and the universe will no longer be subject to sin and decay. At that time the whole of creation will no longer groan under the bondage of sin. If the Restoration Theory is true, then the history of earth is a foreshadow of the history of each individual believer in Jesus Christ. The earth has two beginnings and the Christian has two beginnings. Are there other writers of Scripture who suggest there were two beginnings? Yes, there are.

Let Light Shine

Saved man, the man who accepts Jesus Christ as his Savior, has two beginnings. He has a first beginning when he is born and he has a second beginning when he is "born again." This second birth involves both a washing and a renewing. (Titus 3:5) Something old is destroyed and something new is created. The "old creature" passes away and the "new creature" comes to life. (2 Cor. 5:17) More than any other apostle, Paul constantly taught that sinful man must be restored by God's grace. (Eph. 2:8) He repeatedly returned to this truth of the Scriptures: Fallen man must be restored. Looking at his own background as the great persecutor of Christians, it is no wonder he felt so burdened with this doctrine. It is a recurrent theme and he explained it in many ways and in many places in the New Testament. He often went back to the Old Testament to show how this idea had been demonstrated in Scripture. One such instance was 2 Corinthians 4:6.

2 Cor. 4:6 "For God, who said, 'Let light shine out of darkness,' made his light shine in our hearts to give us the light of the knowledge of the glory of God in the face of Christ." (NIV)

Paul explains the restoration of fallen man by going back to the Genesis account of creation. Paul reminds us that God commanded light to shine on the darkened earth. The parallel he makes is between <u>the darkened state of man</u> and <u>the darkened state of the earth</u>. This parallel is an extremely poor example if the darkness that covered the earth was not the result of sin. If the earth was originally created in darkness, as many anti-Gap Theory people say, then this would mean Paul is teaching that man's darkened state was not the result of sin either. Paul would be teaching that God created Adam that way. Instead of teaching that man "had become" sinful, he would be teaching that man "was" sinful when God first created him. Paul does not teach that! He understands that man was created good, but later became bad. He understands that the earth was created good, but later became bad. This is the only way this comparison makes any sense. He uses God's restoration of the earth as a way of explaining God's restoration of man. Paul understands that Genesis 1:3 was the revelation of light; not the creation of light. How do I know Paul understands this? Paul tells me so. Paul does not say, "For God who CREATED light to shine." Paul says, "For God, who said, 'LET light shine.'" The Greek word he uses is *EPO*, and it doesn't mean to create. It means to speak or call or command. Paul says God commanded the light to shine. He uses this light as a parallel of the light that God shines into our hearts when He saves us. Paul understands that God must first shine His light into fallen man BEFORE he can be restored. His light comes BEFORE our restoration. His light is not created AFTER we believe. Instead, His light existed BEFORE our belief. Although His light exists before we are saved, we cannot see it. It is supernaturally hidden from our fallen hearts just as the light of the sun was supernaturally hidden from the fallen earth. It is not until after God commands His light to shine into our hearts that we can see, "The light of the knowledge of the glory of

God in the face of Jesus Christ." In other words, God has to enlighten us before we can know that when we look at the face of Jesus Christ, we are looking at God Himself. His light must first shine in our hearts before we can attain that knowledge. The light of the Gospel has to be shone into us before we can see the truth of the Gospel. (We must know the truth before we can know the truth.) The light of the Gospel is faith. God must give us the faith to believe the Gospel before we can believe the Gospel. It's not possible for natural man, fallen man, unrestored man to understand the things of God. (1 Cor. 2:14) That's why faith is a gift. (Eph. 2:8) God causes His light to shine into our darkened souls. Paul understands this, so he uses the example of God causing light to shine into the darkened earth. Why does Paul use this example? Because it is a perfect parallel. The first thing God does when He begins the process of restoring us is to shine His light into our hearts. We can't understand, or comprehend, or accept the truth of the Gospel without that light. HE HAS TO CAUSE HIS LIGHT TO SHINE INTO US FIRST. We are spiritually dead. Our souls are darkened because we are a fallen race. We are sinners by virtue of The Fall, and light is not in us. We cannot cause His light to shine in us anymore than the earth could have caused the light of the sun to shine on it. The earth was powerless to create physical light. It could only receive it. We are powerless to create spiritual light. We can only receive what God reveals. Paul uses this example because this is exactly what God did to the fallen earth. He began the restoration process by first causing physical light to shine where sin had caused darkness. The earth was dark because of sin. We were dark because of sin. The earth was in physical darkness and we were in spiritual darkness. Both kinds of darkness were the results of sin. Both kinds of darkness came to an end when God caused light to shine into the darkness. In both cases it is the revelation of light, not the creation of light. Physical light preexisted the physical darkness of the earth. Spiritual light preexisted the spiritual darkness of our hearts. Paul believed man had become dark because of sin. Paul believed the earth had become dark because of sin. Paul makes this comparison because he understands the symbolic connection between man and the earth. Both darkened man and the darkened earth needed light from God to be restored.

All Shook Up

There is another biblical clue that shakes up the various creation theories. As we have seen in Jeremiah 4:24 the earth was shaken by God's wrath sometime in the past. The word used is *RAASH*, and it means to shake, to make tremble, or to make afraid. It is an interesting word! It is often used in connection with Divine judgment. Isaiah 13:13, Ezekiel 38:20, Joel 2:10 and Joel 3:16 use *RAASH* to describe what God will do to at the final judgment.

Isa. 13:9-13 "Behold, the day of the LORD is coming, Cruel, with fury and burning anger, To make the land a desolation; And He will exterminate its sinners from it. *{10}* For the stars of heaven and their constellations Will not flash forth their light; The sun will be dark when it rises, And the moon will not shed its light. *{11}* Thus I will punish the world for its evil, And the wicked for their iniquity; I will also put an end to the arrogance of the proud, And abase the haughtiness of the ruthless. *{12}* I will make mortal man scarcer than pure gold, And mankind than the gold of Ophir. *{13}* Therefore I shall make the heavens tremble, And the earth **will be shaken** from its place At the fury of the LORD of hosts In the day of His burning anger." (NASB)

Eze. 38:18-20 "And it shall come to pass at the same time when Gog shall come against the land of Israel, saith the

Lord GOD, that my fury shall come up in my face. *{19}* For in my jealousy and in the fire of my wrath have I spoken, Surely in that day there shall be a great **shaking** in the land of Israel; *{20}* So that the fishes of the sea, and the fowls of the heaven, and the beasts of the field, and all creeping things that creep upon the earth, and all the men that are upon the face of the earth, **shall shake** at my presence, and the mountains shall be thrown down, and the steep places shall fall, and every wall shall fall to the ground." (KJV)

Joel 2:10 "The earth shall quake before them; the heavens **shall tremble**: the sun and the moon shall be dark, and the stars shall withdraw their shining:" (KJV)

Joel 3:13-17 "Put ye in the sickle, for the harvest is ripe: come, get you down; for the press is full, the vats overflow; for their wickedness is great. *{14}* Multitudes, multitudes in the valley of decision: for the day of the LORD is near in the valley of decision. *{15}* The sun and the moon shall be darkened, and the stars shall withdraw their shining. *{16}* The LORD also shall roar out of Zion, and utter his voice from Jerusalem; and the heavens and the earth **shall shake**: but the LORD will be the hope of his people, and the strength of the children of Israel. *{17}* So shall ye know that I am the LORD your God dwelling in Zion, my holy mountain: then shall Jerusalem be holy, and there shall no strangers pass through her any more." (KJV)

RAASH isn't always used to describe judgment at the end time. It is also used to describe the earth shaking at other times. For instance, *RAASH* is used to describe the shaking of Mt. Sinai in the presence of God. (Psalm 68:8) The earth shook, but it wasn't the end of the world. One distinction in the case of Mt. Sinai is that while several passages indicate there was a tremendous thunderstorm and heavy rainfall, they don't use *RAASH* in connection with the heavens. At Mt. Sinai, only the earth shook. There are other words for "shake" and "tremble" that are also used in connection with both ordinary earthquakes and with the final judgment. *RAASH*, however is used to describe the shaking of both the heavens and the earth at the final judgment in three passages. We've already seen *RAASH* in Joel 3:16. The other two verses are Haggai 2:6 and Haggai 2:21. Let's examine Haggai 2:6 first and then look at Haggai 2:21 later. The Prophet Haggai is looking at the future and he says something that really captures my attention. Haggai 2:6 uses *RAASH* in reference to shaking the heavens and the earth at the final judgment, but it adds one comment easily missed if not read carefully.

Hag. 2:6 "This is what the LORD Almighty says: 'In a little while I will **once more** shake the heavens and the earth, the sea and the dry land.'" (NIV)

"I will **ONCE MORE** shake the heavens and the earth." Do you see what God is telling us? To do this once more means that He's done it before. The same Hebrew words used in Haggai 2:6 (*ECHAD OWD*) are used in Exodus 11:1 where God tells Moses that He will once more bring a plague on Egypt.

Exo. 11:1 "And the LORD said unto Moses, Yet will I bring **one** plague *more* upon Pharaoh, and upon Egypt; afterwards he will let you go hence: when he shall let *you* go, he shall surely thrust you out hence altogether." (KJV)

Exo. 11:1 "Now the LORD said to Moses, '**One more** plague I will bring on Pharaoh and on Egypt; after that he will let you go from here. When he lets you go, he will surely drive you out from here completely.'" (NASB)

God used "one more" because He had already sent plagues on Egypt. The "once more" in Haggai 2:6 means that God had already shaken (judged) the heavens and the earth. Yet, if you search the Scriptures, you'll see no such event after Genesis 1:2. While it is true that the Great Flood of Noah's time was a judgment of the earth, the Great Flood didn't shake the heavens. The Flood went only twenty feet higher than the tallest mountain. The heavens weren't affected. The heavens weren't shaken by The Flood. God had firmly established the heavens, the sun, the moon and the stars to mark the seasons and days and years in Genesis 1:14-18. He set them in place and fixed them in position BEFORE Adam was created. God established their movements with such precision that we can still mark years and seasons and days by their motion. We can still see His invisible attributes in them. The Great Flood of Noah's time was a judgment of the earth, not of the heavens and the earth. God has not shaken (judged) the Post-Adamic heavens and earth. The heavens and earth will not be shaken until the final judgment. "I will **once more** shake the heavens and the earth," can only mean that the Pre-Adamic heavens and earth were shaken by God's judgment. Only the Restoration Theory incorporates a Pre-Adamic judgment of the heavens and the earth.

Now, I know what the Gap Theory doubters will say. Not every translation says, "once more." That's true, but many do. Why do some say, "once more," and others don't? I don't know. I suspect it is grammatically possible to translate it either way. I suspect the translators who left out the "once more" omitted it because they didn't see a need for it. I suppose they thought it might be confusing to say, "once more," since they obviously didn't realize that God had shaken the heavens and the earth once before. Many Gap Theory doubters quote the *King James Version* to prove that it doesn't say, "once more."

Hag. 2:6 "For thus saith the LORD of hosts; **Yet once**, it *is* a little while, and I will shake the heavens, and the earth, and the sea, and the dry *land;*" (KJV)

It would be easy to refute their argument simply by listing the translations that do render it as, "once more" or "once again."

Hag. 2:6 "For thus says the LORD of hosts, 'Once more in a little while, I am going to shake the heavens and the earth, the sea also and the dry land.'" (NASB)

Hag. 2:6 "For thus says the LORD of hosts: Once again, in a little while, I will shake the heavens and the earth and the sea and the dry land;" (NRSV)

Hag. 2:6 "This is what the LORD All-Powerful says: 'In a short time I will once again shake the heavens and the earth, the sea and the dry land.'" (EB)

Hag. 2:6 "For thus says the LORD of hosts: Yet once more, in a little while, I will shake the heavens and the earth and the sea and the dry land." (ESV)

Hag. 2:6 "For thus said יהוה of hosts, 'Once more, in a little while, and I am shaking the heavens and earth, the sea and dry land." (SCRIP)

Hag. 2:6 "For thus said Jehovah of Hosts: Yet once more — it *is* a little, And I am shaking the heavens and the earth, And the sea, and the dry land," (YLT)

Since these other qualified Hebrew scholars translated it, "once more" or "once again," then I believe I have a valid argument. Nevertheless, I want to show those who quote the

King James that this was exactly what the translators were thinking when they said, "yet once."

The Hebrew word for "once" is *ECHAD* and there is no disagreement what this word means. We have seen it before in Genesis 1:5 as *ECHAD YOWM* where they are translated "One Day," "First Day," or "Day One." *ECHAD* is the word for "one," "once," "first," and other such words that derive their meaning from the number one. What we must do, is determine what *OWD* means. Does it mean, "YET?" Does it mean, "AGAIN?" Does it mean, "MORE?" The truth is that it means all of those because they all can have a common meaning. Look at how the King James translators translated *OWD* in other parts of Genesis.

Gen. 4:25 "And Adam knew his wife **again**; and she bare a son, and called his name Seth: For God, said she, hath appointed me another seed instead of Abel, whom Cain slew."

Gen. 8:12 "And he stayed **yet** other seven days; and sent forth the dove; which returned not again unto him any more."

Gen. 9:15 "And I will remember my covenant, which is between me and you and every living creature of all flesh; and the waters shall no **more** become a flood to destroy all flesh."

Gen. 24:20 "And she hasted, and emptied her pitcher into the trough, and ran **again** unto the well to draw water, and drew for all his camels."

Gen. 31:14 "And Rachel and Leah answered and said unto him, Is there **yet** any portion or inheritance for us in our father's house?"

The word "yet" means the same thing as "again" or "more" in this context. If I said, "I have YET to finish writing my book," it means I have MORE of my book to write. The translators of the King James were quite correct to translate it "yet" because they were thinking in sense that there will be one yet (one more) judgment of the heavens and the earth. I'm not sure how you can interpret the words *ECHAD* (one or once) and *OWD* (more or again) except as, "once more." I'm sure Hebrew scholars will argue about this from now until the heavens and the earth are shaken on the last day. I wish them luck! For me, however, the best translation of God's Word is when God Himself interprets what He has written. God does just that. Haggai's words are echoed in the Book of Hebrews.

Heb. 12:25-27 "See to it that you do not refuse Him who is speaking. For if those did not escape when they refused him who warned them on earth, much less shall we escape who turn away from Him who warns from heaven. {26} And His voice shook the earth then, but now He has promised, saying, 'YET ONCE MORE I WILL SHAKE NOT ONLY THE EARTH, BUT ALSO THE HEAVEN.' {27} And this expression, 'Yet once more,' denotes the removing of those things which can be shaken, as of created things, in order that those things which cannot be shaken may remain." (NASB) (emphasis NASB)

The context of this portion of the Book of Hebrews is a warning about the last judgment. At the last judgment both heaven and earth will be shaken. God uses this warning about the last judgment as a way to warn us that our present-day disobedience will lead to present-day judgment. Chapter twelve explains how God disciplines us in order to perfect us. If we don't heed the warnings of His discipline, we will not escape the judgment of His wrath. The writer of Hebrews reminds us that we must be perfect and holy if we are to come to Mount Zion, the heavenly Jerusalem, and dwell in God's presence. Those things which are imperfect and unholy cannot stand before God, and they will be removed from His Kingdom. Two examples of God's judgment are given. The writer of Hebrews reviews what happened to Esau. He sold his birthright, (rejected God) and for that he was rejected by God. The writer also refers to a time when Israel rejected God. God brought the Israelites to Mt. Sinai (Exodus 19) to give Moses the Ten Commandments. God made His presence known; He gave ample proof that it was He who came down to the mountain. At that time, the earth shook and it terrified the Israelites. It terrified Moses too, but he was allowed to approach God. While Moses was away, even though the people knew it was God on the mountain, they rebelled and engaged in pagan orgies and calf worship. They rejected God. They did not heed His warnings, and so they were judged. The writer of Hebrews is urging us not to make the same mistake because God will do something even greater than Sinai when He comes for the final judgment. This won't be just another Sinai. At Sinai, He shook only the earth, but now God tells us, "Yet once more I will shake not only the earth, but also the heaven." The writer of the Book of Hebrews refers back to the words of Haggai. The Holy Spirit inspired both writers to make it clear that He is not talking about a judgment of the earth alone. **The writer of Hebrews verifies that "once more" should be included in the translation of Haggai 2:6.** Scripture clarifies Scripture! Haggai 2:6 should be translated, "In a little while I will **once more** shake the heavens and the earth." The "once more" refers to God shaking "created things," the creation, the physical universe. This is not a shaking of just the earth; this is a shaking of all creation. "Once more" God will judge heavens and earth. The heavens were and will be "shaken." The writer of the Book of Hebrews confirms what

Haggai told us. God has ONCE BEFORE shaken not only the earth, but also the heavens. The heavens and the earth are the physical realm, the "created things," "those things that can be shaken." God has ONCE BEFORE shaken the physical realm. God has ONCE BEFORE judged creation. The Bible tells us that the final judgment of the heavens and the earth is not the first judgment of the heavens and the earth; they have been judged before. Now, if they have been judged before, for whose rebellion were they judged? Not Adam's; he wasn't created yet. Whose sin resulted in God's shaking the heavens and the earth the first time? It was not man. Only the angelic rebellion led by Lucifer could have been the cause.

Before we leave *RAASH* behind, we need to look at Haggai 2:20-23. This is the third passage of Scripture that uses *RAASH* to describe the shaking of both the heavens and the earth. Now, at first reading it seems as if this portion of Scripture is not describing the end times. Instead, it is talking about Zerubbabel, who was the governor of Judah at that time.

Hag. 2:20-23 "And again the word of the LORD came unto Haggai in the four and twentieth day of the month, saying, {21} Speak to Zerubbabel, governor of Judah, saying, **I will shake the heavens and the earth**; {22} And I will overthrow the throne of kingdoms, and I will destroy the strength of the kingdoms of the heathen; and I will overthrow the chariots, and those that ride in them; and the horses and their riders shall come down, every one by the sword of his brother. {23} In that day, saith the LORD of hosts, will I take thee, O Zerubbabel, my servant, the son of Shealtiel, saith the LORD, and will make thee as a signet: for I have chosen thee, saith the LORD of hosts." (KJV)

Zerubbabel was the son of Shealtiel, the son of Jehoiachin, the last legal king of Judah before Babylon destroyed

Jerusalem. Zerubbabel was heir to the throne and he should
have been Judah's king. Here we see that he is only its governor. Why wasn't he the king? We need to see what happened to dear Grandpa Jehoiachin. King Jehoiachin was the
last of King David's royal line to sit on the throne legally. I
say, "legally," because Jehoiachin was carried off to Babylon,
and Babylon put Jehoiachin's uncle Zedekiah on the throne.
Jehoiachin's son Shealtiel had the legal claim to the throne,
but Babylon chose its own king to rule in Jerusalem instead.
It may have been that Shealtiel wasn't old enough, but it was
Babylon, not God, who made Zedekiah king. Zedekiah had
no legal claim to David's throne in God's eyes. (Zedekiah
later rebelled against Babylon and was carried off there to
die.) Jehoiachin, therefore, was Israel's last legally reigning
king before the Babylonian conquest. Now, Jehoiachin had
two other names in the Bible. He was called Jeconiah, and
he was also called Coniah. He was an evil king, and the Lord
brought judgment on him when Babylon captured Judah and
destroyed Jerusalem. Coniah never returned from Babylon.
Coniah was so evil, that God cursed him and declared that
none of his offspring would ever be king.

Jer. 22:24-30 "As I live, saith Jehovah, though Coniah the
son of Jehoiakim king of Judah were the signet upon my
right hand, yet would I pluck thee thence; {25} and I will
give thee into the hand of them that seek thy life, and into
the hand of them of whom thou art afraid, even into the hand
of Nebuchadrezzar king of Babylon, and into the hand of
the Chaldeans. {26} And I will cast thee out, and thy mother
that bare thee, into another country, where ye were not born;
and there shall ye die. {27} But to the land whereunto their
soul longeth to return, thither shall they not return. {28} Is
this man Coniah a despised broken vessel? is he a vessel
wherein none delighteth? wherefore are they cast out, he and
his seed, and are cast into the land which they know not?

{29} O earth, earth, earth, hear the word of Jehovah. {30} Thus saith Jehovah, Write ye this man childless, a man that shall not prosper in his days; for no more shall a man of his seed prosper, sitting upon the throne of David, and ruling in Judah." (KJV)

Several years after Babylon conquered Israel, Babylon itself was conquered by the Persians. King Cyrus of Persia made a decree to have Jerusalem rebuilt, and he sent Zerubbabel to be its governor. Was Haggai looking at the near future? Was he anticipating that God would restore Zerubbabel to the throne? No, Haggai was not looking at the immediate future of Zerubbabel. Haggai was looking much farther into the future. Haggai 2:21 indicates that he was looking to the time when God was going to shake the heavens and the earth, the final judgment. How do I know that Haggai wasn't looking at the future of Zerubbabel himself? The answer is because this never happened in Zerubbabel's time. Zerubbabel was never made king. He wasn't restored to David's Throne. All of the events that Haggai predicted haven't yet come to pass. Did God break His promise? No, this passage does not say that God would make Zerubbabel the king. Instead, it says that He would make him a "signet ring." What did this mean?

Look back at what God said about Coniah. If Coniah was a signet ring on his right hand, He would pluck him off. A signet ring was a symbol of legal authority. In other words, God removed Coniah's legal claim to the throne. No one of Coniah's bloodline would ever be King of Israel. Now look at what God said about Zerubbabel, the grandson of Coniah. God said He would make Zerubbabel a signet ring. This indicated that God would give the legal claim to the throne back to Zerubbabel. That was great, but the curse of Coniah remained; Zerubbabel never became king. Haggai wasn't looking at Zerubbabel. He wasn't looking at

Zerubbabel's son. He didn't become king either. Nor was he looking at Zerubbabel's grandson, great-grandson, great-great-grandson, or great-great-great-grandson. Even though they all held the legal claim to David's Throne, none of Zerubbabel's offspring became King because they were of Coniah's bloodline. They were the Kings in Exile.

Ten generations later (Matthew 1:13-16) the last King in Exile was born. His name was Joseph, and he was a carpenter who lived in Nazareth. He was betrothed to a young virgin named Mary. By the power of the Holy Spirit, Mary conceived a son named Jesus. Joseph was not the biologic father of Jesus, but by marriage and adoption, Joseph was the legal father of Jesus. That meant that Jesus held the legal claim to the throne of David. At the same time, Jesus was not of Coniah's bloodline. The curse of Coniah did not apply to Him. He had the right to claim the Throne of David and establish Himself as King of Israel, but He didn't do it. Oh, the crowds wanted Him to. His disciples wanted Him to. Even Satan tried to get Him to take the throne, but Jesus refused. He refused because He knew He was not to set up His kingdom until the heavens and earth were judged in the future. Evil would be removed from His Kingdom before He took the Throne. The promise that God made to Zerubbabel was fulfilled. Zerubbabel wasn't made King, but the "signet," the authority of the Royal Line, was continued through him. In the end, his greatest Son, Jesus, will become the King. Haggai 2:21 is referring to a future judgment of the heavens and the earth. Haggai looked forward to the time when the heavens and the earth will be shaken. But as he did so, he also looked back at the time when they had been shaken before. That's why Haggai said, "In a little while I will once more shake the heavens and the earth, the sea and the dry land." Haggai knew the earth had two beginnings.

Two Global Floods

According to creationists who reject the Gap Theory, there was only one beginning. They don't believe the earth had an original creation, a destruction, and a later restoration. They don't believe the earth had two beginnings. So, they interpret Genesis 1:1-2 as being the first part of day one. This means that God created the earth in the condition that Genesis 1:2 describes it. What was that condition? It was dark and covered by water. Young-Earth creationists believe that God initially created the earth that way. They think Genesis 1:1-2 was a very short period of the first twenty-four hour day. They believe that God created the earth already dark and already covered by water. Being dark and covered by water was not something the earth **came to be**; it was something the earth **was** when it was created. This is their interpretation of *HAYAH*. The earth was dark and the earth was covered by water when God first created it. Light and dry land had not yet been created according to Young-Earth creationists. Day-Agers, on the other hand, realize the earth couldn't have been created that way. Day-Age creationists know the scientific evidence doesn't support the idea that the earth was covered by water for billions of years before the crust (dry land) was formed. They also know the scientific evidence doesn't support the idea that the earth was present for billions of years before the sun and the stars came into existence. Day-Age creationists know the sun and the stars were already shining when the earth was created. The earth did not begin in darkness. They know the universe and the earth are billions of years old, so they interpret the "days" of Genesis as being long ages. Day-Age creationists think Genesis 1:1 was a very long period of time. The first "day" was billions of years. But, as we have seen, this creates a problem with the order of appearance of things. If the days are long ages, and if they are in the correct order, then the earth was dark and covered by water for billions of years before the sun, moon, and stars were visible. Day one has to come before day four or else God made a mistake in His account of creation in Genesis. Day-Agers reject our interpretation of *HAYAH*, but they can't reconcile the order of creation unless our interpretation of *HAYAH* is correct. Day-Agers have a real problem if you ask them, "When was Genesis 1:2? When was the earth completely covered by water (global flood) and in complete darkness?" They know there was light before the earth was created. It didn't begin in darkness. Somehow, it must have **become** dark after its original beginning. They know the earth's hot crust was present long before surface water formed. The earth did not begin under water. Somehow, it must have **become** covered by water after its original beginning. They know the earth had to **become** dark and **become** covered by water, but they still insist that we Gap Theory creationists are wrong to say that *HAYAH* means "**had become.**" We Gap Theory creationists believe the earth **became** covered by water. We Gap Theory creationists believe there was light before the earth **became** dark. We Gap Theory creationists believe the earth **became** dark and covered by water sometime after its original creation. Generally, we Gap Theory creationists believe that God judged the original earth (covered it in darkness) and destroyed the life on it by causing a worldwide flood. In other words, the earth has experienced two global floods. One was at the time of Noah and the other was before Adam's creation. **The fact that the earth has been subjected to two global floods is a foundational component of the Gap Theory.** Both Young-Earth and Day-Age creationists reject the idea that earth has been flooded twice. That's a shame because God apparently believes the world was flooded twice. Look again at what He told Job.

Job 38:1-11 "Then the LORD answered Job out of the whirlwind and said, {2} 'Who is this that darkens counsel By words without knowledge? {3} 'Now gird up your loins like a man, And I will ask you, and you instruct Me! {4} 'Where were you when I laid the foundation of the earth? Tell *Me,* if you have understanding, {5} Who set its measurements, since you know? Or who stretched the line on it? {6} 'On what were its bases sunk? Or who laid its cornerstone, {7} When the morning stars sang together, And all the sons of God shouted for joy? {8} 'Or *who* enclosed the sea with doors, When, bursting forth, it went out from the womb; {9} When I made a cloud its garment, And thick darkness its swaddling band, {10} And I placed boundaries on it, And I set a bolt and doors, {11} And I said, 'Thus far you shall come, but no farther; And here shall your proud waves stop'?" (NASB)

Look very closely at what God reveals to Job about this global flood. God tells Job there was a time when the sea burst forth like a pregnant woman's "water" breaking just before she gives birth. Was this the Great Flood of Noah's time? No, God is giving Job details about the creation of the earth. He is questioning Job about his knowledge of what happened at the creation. The imagery of a pregnant woman's "water" breaking is an imagery of birth. This event is part of the birthing process; it begins the birth. This birthing imagery sets the time frame to be that of earth's beginning. (Actually, the earth's second beginning; Gap Theory creationists believe that Job 38:1-7 refer to earth's first beginning, while Job 38:8-11 refer to earth's second beginning. These passages of Scripture are separated by the word "or.") But, didn't the waters burst forth at the time of Noah's flood? Yes, both Job 38:8 and Genesis 7:11 describe the waters bursting forth, but they describe different global floods. These aren't the same flood because the details don't match.

In Job, God says the earth was covered by water and **thick darkness**. The earth was dark during this flood, but the earth wasn't dark during Noah's Flood. In Noah's time, the waters came upon the earth for "forty days and forty nights." There were days and nights. There was light during Noah's Flood; the earth wasn't dark. In addition, God tells Job that these waters **didn't completely cover the earth**. God tells Job He, "enclosed the sea with doors." He placed boundaries on it. He stopped its proud waves. They reached a certain point and He let them go no farther. This wasn't true at Noah's time. God did not stop the waves from covering the entire earth during Noah's Flood. All the earth was covered, even the tallest mountains. God is telling Job about a different global flood. God is describing the earth at a different time. In Noah's Flood, the entire earth **was** covered by water. In Noah's Flood there was **light**. In this other global flood, the entire earth **wasn't** covered by water. In this other global flood, there was **no light**. These are two different global floods. The earth has been flooded twice!

Global Flood 1	**Global Flood 2 (Noah's Flood)**
Earth covered in darkness	Earth not covered in darkness
Earth not completely covered by water	Earth completely covered by water

When was this other global flood? God told Job that He flooded the earth, "When I made a cloud its garment, and thick darkness its swaddling band." In other words, God first made the earth dark, and then He flooded it. This wasn't done at the time of Noah. God didn't make the earth dark. It couldn't be Noah's Flood. It couldn't have been AFTER Noah's Flood either. God promised He would never flood the earth again. It can't be in the future for the very same reason. It couldn't have been between the time of Adam and the time of Noah because the earth wasn't dark during that time. The sun, moon, and stars were shining and set in the firmament of heaven to mark days, years, and seasons even before Adam was created. There was never a time between Adam and Noah that the earth was dark and covered by water. God is telling Job about a global flood that happened before Adam was created. God is telling Job about a Pre-Adamic global flood. God says the waters burst forth. The earth **became** covered by water only **after** the waters burst forth. The water that covered the earth in Job 38 burst forth from the earth (dry land) that was already there. The earth **became** covered by water; it didn't begin that way. There was dry land before the earth was flooded. It was only after the waters burst forth that the earth became covered by water. In the same way, God **covered** the earth in darkness. It wasn't created that way. Just as a baby is born before it is wrapped in swaddling bands, the earth was created before it was wrapped in darkness. Again, God uses the imagery of birth. The earth was not "born" in thick darkness. God wrapped it in thick darkness (judgment) sometime after it was created. The earth was created in light; then it later **became** dark. If you recall Jeremiah's description of the early earth, you will see that God again reveals the earth in a condition that doesn't exactly match the condition of the earth in Genesis 1:2.

Jer. 4:23-26 KJV "I beheld the earth, and, lo, *it was* without form, and void; and the heavens, and they *had* no light. *{24}* I beheld the mountains, and, lo, they trembled, and all the hills moved lightly. *{25}* I beheld, and, lo, *there was* no man, and all the birds of the heavens were fled. *{26}* I beheld, and, lo, the fruitful place *was* a wilderness, and all the cities thereof were broken down at the presence of the LORD, *and* by his fierce anger." (KJV)

In Genesis 1:2 the earth is dark and covered by water. There is no dry land and there is no light. It's not until Genesis 1:3 that God creates (restores) light. In Genesis 1:3 there is light, but no dry land. Jeremiah, on the other hand, describes earth at a time when there is dry land but no light. He beheld mountains, hills, fruitful places, wilderness, etc., but the heavens had no light. **GENESIS ONE HAS LIGHT BEFORE LAND, BUT JEREMIAH FOUR HAS LAND BEFORE LIGHT. GENESIS ONE HAS LIGHT BEFORE LAND, BUT JEREMIAH FOUR HAS LAND BEFORE LIGHT.** Jeremiah and Moses could not both be describing earth at the same time unless the Bible contradicts itself. God has given us two accounts of earth's condition in the past. These accounts don't agree with each other concerning the appearance of light and of dry land. God has made a mistake in the Bible if there was only one beginning. My friends, God makes no mistakes. He has given us the accounts of different time periods in the Pre-Adamic earth. Moses reveals that God created the earth. Jeremiah, Job, and Haggai reveal that God brought darkness upon the earth, judged (shook) the earth, and caused water to burst forth and cover it. It was at that point that the earth became formless and void, with darkness covering the face of the deep. It wasn't that way before. God did not create the earth that way. (Isa. 45:18) He tells us that this was what the earth BECAME sometime after its original creation. Why don't we believe Him?

The Past Age

Heb. 11:3 "Through faith we understand that the worlds were framed by the word of God, so that things which are seen were not made of things which do appear." (KJV)

Hebrews 11:3 has long been a verse used to defend the Gap Theory. This is because the word for "framed" has a variety of meanings. Some of those meanings can be taken as evidence that the earth was destroyed and then restored. The word is *KATARTIDZO*. Its meanings include "to prepare," "to adjust," "to fit," "to frame," to perfect," "to repair," "to restore," "to mend," and similar things. Let me give you an example of how it is used in the sense of "to repair."

Mat. 4:21 "And going on from thence, he saw other two brethren, James *the son* of Zebedee, and John his brother, in a ship with Zebedee their father, **mending** their nets; and he called to them." (KJV)

Certainly the fact that they were mending their nets indicated that their nets were damaged and needed repair. By applying this idea to the creation of the world, it could be argued that the world was "repaired" by the Word of God. I can't argue that it isn't possible, and it may be what God is telling us, but I have to agree with those who disagree. *KATARTIDZO* doesn't always mean to repair something that is broken or in ruin. That is one of its most common meanings, but to insist that it must mean "repair" is too close to the edge of blasphemy for me.

Heb. 10:5 "Wherefore when he cometh into the world, he saith, Sacrifice and offering thou wouldest not, but a body hast thou **prepared** me." (KJV)

I don't think I want to defend a doctrine preaching that Christ's body was fallen, damaged, in ruin, or in need of repair. He was the perfect, spotless Lamb, without blemish or fault. If He had any defect, then He could not have offered Himself as the perfect Sacrifice who atoned for our sins. So, while this definition of *KATARTIDZO* may defend the Gap

Theory, I don't necessarily want to defend it. I do want to focus your attention on something else. Read Hebrews 11:3 again. What was framed? "The worlds" were framed. It's in the plural. How many worlds have we had? We can't answer that until we know what "worlds" means. Some interpret Hebrews 11:3 to mean all the planets and galaxies, and so translate it, "the universe."

Heb. 11:3 "By faith we understand that the **universe** was formed at God's command, so that what is seen was not made out of what was visible." (NIV)

The word is *AION*, from which we get "eon." It means "a period of time" or "an age."

Heb. 11:3 "By faith we understand the **ages** to have been prepared by a saying of God, in regard to the things seen not having come out of things appearing;" (YLT)

If I were a Day-Ager, I would insist that this was proof for the Day-Age Theory, and I would jump up and down in excitement. But I'm not, and it isn't, so I won't. Being a Gap Theory creationist, this verse excites me too, but for a different reason. These "ages" were "framed" in the past. According to *A Parsing Guide to the New Testament*,[44] the verb *KATARTIDZO* (framed) in Hebrews 11:3 is in the perfect, passive, infinitive form. I'm not much better with Koine Greek than I am with ancient Hebrew, but according to Greek scholars, this means it is a verbal noun with action on *AION* that was completed in the past but with finished results that remain into the present. In other words, these "ages" were completed in the past. Let's look at what this means. *AION* means an age, but it means more than just a measure of time like "century" or "millennia." It incorporates not only a passage of time; it incorporates the events and the conditions that occupy that passage of time. That's why we can talk about such things as "The Stone Age," "The Iron Age," "The Industrial Age," "The Age of Enlightenment." These designations don't define specific periods of time such as "The Fifth Century B. C." or our present "Twenty-First Century." Yet incorporated into *AION*, we do have the concept of specific periods of time. There was a specific period of time, a long time ago, that stone tools and weapons were the height of man's technology. This was the "Time-Space" period of "The Stone Age." If I was talking about "The Computer Age," even though I didn't give you specifics dates, you would know that I wasn't talking about the "Time-Space" period of the Egyptian Pharaohs. *AION* combines some aspects of time with some aspects of space. (Since time and space are a continuum, I think the Holy Spirit did a smashing good job at picking a word that incorporates aspects of both concepts.) Because it incorporates both concepts, it is often translated "world."

Mat. 13:22 "And the one on whom seed was sown among the thorns, this is the man who hears the word, and the worry of the **world**, and the deceitfulness of riches choke the word, and it becomes unfruitful." (NASB)

Here are some more verses that translate *AION* (in bold print) as "age."

Mat. 12:32 "And whoever shall speak a word against the Son of Man, it shall be forgiven him; but whoever shall speak against the Holy Spirit, it shall not be forgiven him, either in this **age**, or in the *age* to come." (NASB)

Mat. 13:39 "and the enemy who sowed them is the devil, and the harvest is the end of the **age**; and the reapers are angels." (NASB)

Mat. 13:40 "Therefore just as the tares are gathered up and burned with fire, so shall it be at the end of the **age**." (NASB)

Mat. 13:49 "So it will be at the end of the **age**; the angels shall come forth, and take out the wicked from among the righteous," (NASB)

Mat. 24:3 "And as He was sitting on the Mount of Olives, the disciples came to Him privately, saying, "Tell us, when will these things be, and what *will be* the sign of Your coming, and of the end of the **age**?" (NASB)

Eph. 1:18-21 "*I pray that* the eyes of your heart may be enlightened, so that you may know what is the hope of His calling, what are the riches of the glory of His inheritance in the saints, *{19}* and what is the surpassing greatness of His power toward us who believe. *These are* in accordance with the working of the strength of His might *{20}* which He brought about in Christ, when He raised Him from the dead, and seated Him at His right hand in the heavenly *places, {21}* far above all rule and authority and power and dominion, and every name that is named, not only in this **age**, but also in the one to come." (NASB)

From these and many other passages we see that there is a present age and there is an age to come. There is a "this age" and a "that age." That age, the age that comes after this age, is still future. It hasn't been made yet. That future age is not included in the ages mentioned in Hebrews 11:3 because those ages were prepared in the past. The verb form indicates that those ages were completed in the past, but still have results that remain in the present. The age to come has not been completed and its results are not yet present. I'm sure you see where I'm going with this. Since there were ages completed in the past, and since we are in the present age now, then the Bible reveals that there must have been an age before this age.

I'd like to end my argument right here, but I see its own weakness. There could be another way of looking at this. What if we divided earth's history into "ages" such as "the Age of Innocence," "the Antediluvian Age," "the Age of the Patriarchs," "the Jewish Age," and "the Church Age?" If these were the "ages" that were "framed," then this is not a defense for the Gap Theory. There wouldn't need to be a past age before the present age to satisfy the grammatical qualities of Hebrews 11:3. If we can learn what period of time the Bible ascribes to the present age, then we can see if Hebrews 11:3 reveals anything significant. We need to seek a biblical definition for "this age." Seek and ye shall find.

Luke 20:34-35 "And Jesus said to them, 'The sons of **this age** marry and are given in marriage, *{35}* but those who are considered worthy to attain to **that age** and the resurrection from the dead, neither marry, nor are given in marriage;'" (NASB)

Jesus defines THIS AGE as the age when people marry. When did marriage start? It started with Adam and Eve. This means Adam and Eve were part of this age. When will people no longer marry? It will be after this age has ended. It will be in the age to come. Jesus defines the present age as the time from the creation of Adam and Eve until the time of the resurrection of the dead. **Jesus tells us that there has been only one age since the creation of Adam and Eve.** Hebrews 11:3 refers to a past age. So if there has been only one age since the creation of Adam and Eve, then Hebrews 11:3 is telling us there was an age before the creation of Adam and Eve. That previous *AION* was a period of time before this present *AION*. There was an age before this age. How long was that age? I don't know, but I certainly wouldn't feel comfortable in calling five days an age. Day One through Five of Genesis 1 is the only measurable period of time before Adam that Young-Earth creationists allow. Calling five days "an age" doesn't make much sense. There must have been a long measurable period of time prior to Adam. That's not intended to mean that there was just a measurable period of time before this period of time. Remember that the word *AION* includes more than just the passage of time; it includes the events and the conditions that occupy that passage of time. *AION* in Hebrews 11:3 seems to indicate there were events and conditions that were happening in the age before the events and conditions of this age. Earth has experienced two different events and conditions; two different ages. The earth has had two beginnings.

The Gap Theory Under The Electron Microscope

I want you to read the following passages of Scripture because I'm going to give you a quiz. Pay close attention to the word "foundation." Here is the quiz: After reading these passages from God's Word, what do you think the Greek word *THEMELIOS* (the bold print) means?

Luke 6:48-49 "He is like a man which built an house, and digged deep, and laid the **foundation** on a rock: and when the flood arose, the stream beat vehemently upon that house, and could not shake it: for it was founded upon a rock. {49} But he that heareth, and doeth not, is like a man that without a **foundation** built an house upon the earth; against which the stream did beat vehemently, and immediately it fell; and the ruin of that house was great."

Luke 14:29 "Lest haply, after he hath laid the **foundation**, and is not able to finish *it*, all that behold *it* begin to mock him,"

Acts 16:26 "And suddenly there was a great earthquake, so that the **foundations** of the prison were shaken: and immediately all the doors were opened, and every one's bands were loosed."

Rom. 15:20 "Yea, so have I strived to preach the gospel, not where Christ was named, lest I should build upon another man's **foundation**:"

1 Cor. 3:10-12 "According to the grace of God which is given unto me, as a wise masterbuilder, I have laid the **foundation**, and another buildeth thereon. But let every man take heed how he buildeth thereupon. *{11}* For other **foundation** can no man lay than that is laid, which is Jesus Christ. *{12}* Now if any man build upon this **foundation** gold, silver, precious stones, wood, hay, stubble;"

Eph. 2:20 "And are built upon the **foundation** of the apostles and prophets, Jesus Christ himself being the chief corner *stone;*"

1 Tim. 6:19 "Laying up in store for themselves a good **foundation** against the time to come, that they may lay hold on eternal life."

2 Tim. 2:19 "Nevertheless the **foundation** of God standeth sure, having this seal, The Lord knoweth them that are his. And, Let every one that nameth the name of Christ depart from iniquity."

Heb. 6:1 "Therefore leaving the principles of the doctrine of Christ, let us go on unto perfection; not laying again the **foundation** of repentance from dead works, and of faith toward God,"

Heb. 11:10 "For he looked for a city which hath **foundations**, whose builder and maker *is* God."

Rev. 21:14 "And the wall of the city had twelve **foundations**, and in them the names of the twelve apostles of the Lamb."

Rev. 21:19 "And the **foundations** of the wall of the city *were* garnished with all manner of precious stones. The first **foundation** *was* jasper; the second, sapphire; the third, a chalcedony; the fourth, an emerald;"

Do you have an answer to my quiz? Of course you do! It means "foundation." A foundation is something you start with when you want to build something. If we're talking about a building, a foundation is a solid structure you lay down, upon which you erect the rest of the building. If we're talking about an idea, a philosophy, or a theology, then a foundation is a basic concept, (or concepts) upon which the rest of its teachings are established. If that was your answer, then you get an A+. That is exactly what *THEMELIOS* means.

Now let me give you another quiz. Again, look at the word "foundation" (in bold print) and tell me what you think the Greek word means in these passages of Scripture.

Mat. 13:35 "That it might be fulfilled which was spoken by the prophet, saying, I will open my mouth in parables; I will utter things which have been kept secret from the **foundation** of the world."

Mat. 25:34 "Then shall the King say unto them on his right hand, Come, ye blessed of my Father, inherit the kingdom prepared for you from the **foundation** of the world:"

Luke 11:50 "That the blood of all the prophets, which was shed from the **foundation** of the world, may be required of this generation;"

John 17:24 "Father, I will that they also, whom thou hast given me, be with me where I am; that they may behold my glory, which thou hast given me: for thou lovedst me before the **foundation** of the world."

Eph. 1:4 "According as he hath chosen us in him before the **foundation** of the world, that we should be holy and without blame before him in love:"

Heb. 4:3 "For we which have believed do enter into rest, as he said, As I have sworn in my wrath, if they shall enter into my rest: although the works were finished from the **foundation** of the world."

Heb. 9:26 "For then must he often have suffered since the **foundation** of the world: but now once in the end of the world hath he appeared to put away sin by the sacrifice of himself."

1 Pet. 1:20 "Who verily was foreordained before the **foundation** of the world, but was manifest in these last times for you,"

Rev. 13:8 "And all that dwell upon the earth shall worship him, whose names are not written in the book of life of the Lamb slain from the **foundation** of the world."

Rev. 17:8 "The beast that thou sawest was, and is not; and shall ascend out of the bottomless pit, and go into perdition: and they that dwell on the earth shall wonder, whose names were not written in the book of life from the **foundation** of the world, when they behold the beast that was, and is not, and yet is."

Okay, if you have an answer that is similar to the answer in the first quiz, then I have some good news for you and some bad news for you. The good news is that you answered it the same way virtually every Greek Bible scholar would answer it. The bad news is that you all get an F- for giving that answer. Oh boy, I'm in trouble now. I can just hear the scholars criticizing me for my ignorance and my arrogance. Before I tell you what the Greek word is, I must tell you that I have decided to translate it in a way that very few scholars would accept. Even the scholar I most admire, Arthur Custance,[45] explained how he had once translated it this way and felt satisfied with it, but then came to feel he didn't quite have the linguistic support to defend it. I think Custance would have stuck to his original inclination if he had gone to his local biologist and asked for the electron microscope version of "the birds and the bees."

I'm sure you're confused by now, or else you think I am. Here is how this all comes together. The Greek word used in this second set of passages is not *THEMELIOS*. *THEMELIOS* is the usual word for "foundation." The word used in these passages is *KATABOLE*. *KATABOLE* is the word from which we get "catabolic," "catabolism," and "catabolize." Any high school biology student would instantly recognize the significance of this word. It means to break down, tear apart, or destroy. Biologically, it defines that portion of metabolism in which large or complex structures are broken down to be used for energy or for building new structures via anabolism. Old proteins are catabolized into amino acids so they can be anabolized into new proteins. Carbohydrates and fats undergo similar "destruction" so their components can be used for any number of biological purposes. Since *KATABOLE* is combined with "world"

(*KOSMOS*) in these verses, one immediately wonders what and when was the *KATABOLE KOSMOS*, the catabolism of the world? When was the world broken down, destroyed, or torn apart? The context seems to be the period of time that is elsewhere described as the beginning (*ARCHE*) of the earth. **I believe if I can show that *KATABOLE* means "destruction" when used in connection to the beginning of the earth, then I think I have proven the Gap Theory.** As you probably suspect though, this isn't as clear cut as it may seem. How do you get "foundation" out of a word that seems to mean a destructive process?

As I mentioned, Arthur Custance once favored this translation as a defense of the Gap Theory. He suggested that it should be translated "disruption," but he pulled back a little when he couldn't find good linguistic support for it. He didn't reject it altogether, but he declined to argue the point. The problem was that *KATABOLE* was never (or very seldom) translated as a destructive process by Bible scholars. The reason they didn't translate it that way was because... well, because Bible scholars never translated it that way. Custance saw the cyclic reasoning in that argument, but without some other evidence, he knew he couldn't prove that *KATABOLE* meant a destructive process. Certainly, Bible scholars were aware that it COULD define a destructive process, but it wasn't used that way enough to make it a good solid translation. Nobody was willing to champion that definition. Scholars don't like making up unique definitions. Unique definitions are difficult to defend when other scholars attack them. Since I'm not a scholar, I'm not worried about their attacks. I'm only a little boy playing chess.

Sometimes you have to make a unique definition. When a word is used only once or when it is used in only one way, then there is no other choice but to make a unique definition. This was the situation surrounding the definition of *KATABOLE*. The only apparent clue was that *KATABOLE* was a noun derived from the verb *KATABALLO*. The verb *KATABALLO* was easy to define from Greek literature. It meant to throw down, cast down, break down, destroy, tear down, etc. It also meant to lay down, like laying down a foundation. A foundation was built by casting down or laying down or depositing solid material so that something could be built on it. The material used for a foundation was often stones and rubble torn down (catabolized) from older structures. So, *KATABALLO* expressed two distinct, but related, concepts. One was to tear something down, the other was to lay something down or to deposit something. It is used only three times in the Bible.

2 Cor. 4:9 "Persecuted, but not forsaken; **cast down**, but not destroyed;" (KJV)

Heb. 6:1 "Therefore leaving the principles of the doctrine of Christ, let us go on unto perfection; not **laying** again the foundation of repentance from dead works, and of faith toward God," (KJV)

Rev. 12:10 "And I heard a loud voice saying in heaven, Now is come salvation, and strength, and the kingdom of our God, and the power of his Christ: for the accuser of our brethren is **cast down**, which accused them before our God day and night." (KJV)

Note that in 2 Corinthians 4:9, "cast down" (*KATABALLO*) is not the same thing as "destroyed." *APOLLUMI* is the word used for "destroyed." *KATABALLO* means to tear down or break down, but it doesn't necessarily imply an absolute destruction. Rubble wasn't absolutely destroyed when it was used to lay foundations for new buildings. I think this implies that the *KATABOLE KOSMOS* wasn't a total destruction of the earth, but a tearing down or breaking down with the intent of rebuilding something new on it. Now look at Hebrews 6:1. Some people argue that it should say, "Let us go on unto perfection; not **destroying** again the foundation of repentance..." This seems to make more sense, and if it is so, then *KATABALLO* is used only in its destructive sense in the Bible. Still, the evidence wasn't clear enough to insist on either meaning. *KATABALLO* could mean "tearing down" or it could mean "laying down." But, look again at Hebrews 6:1. Even if it is used in the sense of "laying down," it isn't the foundation itself. *THEMELIOS* is still the foundation. *KATABALLO* may be the verb for the laying down of a foundation, but it is not the foundation. This makes me think we need to be careful if we try to translate the noun *KATABOLE* as "foundation" in the Bible. Anyway, here's where things start becoming strange. Somewhere along the way (and I don't even want to know how this came about) it developed a sexual association. When a man deposited semen into a woman, he was "throwing down," "casting down," or "laying down" his seed into her. In essence, he was laying down a "foundation" for a new life. The specific definition of *KATABALLO* was determined by its context. Throughout Greek literature it was probably obvious when a writer was writing about tearing down rubble and when he was writing about men and women doing you know what.

That was the verb *KATABALLO*. What about the noun *KATABOLE*? It was a little harder to define, especially when used in the Bible. Custance noted three problems. The first problem was that nouns don't always mean the exact same thing their related verbs mean. Usually they do, but without linguistic support, no one can use that argument. The second problem with *KATABOLE* in the Bible was that in all of its uses, the context never made it clear which definition was intended. The third problem with *KATABOLE* in the Bible was that it was always used in the same way (with one exception) with the same kind of ambiguity. With only one exception, it was always connected with *KOSMOS*, and what did that mean? What was *KATABOLE KOSMOS*? Whatever definition was used, it would be a unique definition by virtue of the fact that it wasn't used any other way. So, they always translated it the way it had always been translated. Now, if no one really knew how it was supposed to be translated, then always translating it the same way didn't prove that the translation was correct. Making the same mistake ten times doesn't eliminate the fact that a mistake was made. *KATABOLE* was used in eleven verses and in ten of them, it was connected with *KOSMOS*. In all ten of those verses, the Holy Spirit just didn't seem to supply enough context to provide a good definition. So, it was the eleventh verse that scholars looked to in order to find its meaning.

Heb. 11:11 "Through faith also Sara herself received strength to **conceive** seed, and was delivered of a child when she was past age, because she judged him faithful who had promised." (KJV)

The *King James Bible* scholars translated *KATABOLE* as "conceive," as did many other scholars. It was translated as a verb, but actually it was a noun. In fact, two nouns were used: *KATABOLE* and *SPERMA* (seed). Translating it as a verb was no problem; all the scholars agreed. But there was a problem! The problem with defining *KATABOLE* as "conceive" is that it isn't the normal Greek word used for "conceive" in the Bible. The Bible used the word *SULLAMBANO*. It was used when Elizabeth conceived John and when Mary conceived Jesus. If this was what the Holy Spirit meant to say, then why did He pick *KATABOLE SPERMA* and not *SULLAMBANO SPERMA*? What did *KATABOLE SPERMA* mean?

The idea that the noun *KATABOLE* carried any meaning of destruction didn't seem very likely. Sarah was already barren and if God gave her, "power unto the destruction of seed," then she wouldn't have become pregnant with Isaac. Destroying *SPERMA* didn't seem like a good translation. Depositing *SPERMA* or laying down *SPERMA* seemed better. By translating it this way, it was then possible to define *KATABOLE KOSMOS* as "the laying down (the foundation) of the world." This was a linguistic clue: Since Hebrews 11:11 explained how Sarah was able to become pregnant, "destruction" didn't seem right. If "destruction" didn't make sense in Sarah's case, then "destruction of the world" didn't make sense either. If *KATABOLE* really did mean destruction in reference to the earth, then it would also seem to mean that Sarah became able to destroy sperm. The only logical option was to translate *KATABOLE* in the sense of laying down sperm. But that option wasn't without difficulties itself. If *KATABOLE* meant "deposit," then a new problem arose. What did it mean for Sarah to be given the power to deposit *SPERMA*?

This brought up all kinds of opinions about what happens during conception. At that time, no one knew what really happened during the conception event. Some speculated that the man produced the "seed" and the woman was just the "garden" in which the "seed" grew. Others believed that the woman produced the "seed" and the man just provided the "fertilizer" that stimulated the "seed" to grow. Still others believed that "seed" was a figurative term, and it was the commingling of the man's fluids with the woman's fluids that started the process. The fluids congealed and a baby would develop from that little blob of jelly. There were others who were more correct, who believed that both the man and the woman produced "seed." These two "seeds" would combine to make one new "seed." After all, the Bible did talk about the "seed of the woman" and "the seed of the man." (Another pre-scientific revelation of a scientific truth.) But if the text meant that Sarah deposited "seed," then it could be interpreted to mean that Sarah impregnated herself. That couldn't be right. So, rather than using such words as "deposit," seed or "lay down" seed, the word "conceive" seed seem more reasonable. It wasn't accurate, but it was better than "destroy" seed. No matter what the process was, the destruction of seed (either hers or Abraham's or the combination) didn't fit with their knowledge of conception.

Not everyone was happy with that translation, however. There were scholars who insisted that *KATABOLE SPERMA* didn't mean conceive seed. Instead, it meant either deposit seed or destroy seed. They knew it couldn't mean that Sarah was able to destroy seed, and they didn't like the idea of Sarah depositing seed. So they chose another translation. What did these scholars do when faced with the dilemma of Sarah either becoming empowered to destroy seed or becoming empowered to deposit seed? Neither translation made any sense, so they decided God must have meant that Abraham was given the power to deposit seed.

Heb. 11:11 "By faith **Abraham**, even though **he** was past age— and Sarah herself was barren— was able to become a father because he considered him faithful who made the promise." (NIV)

Heb. 11:11 "By faith **he** received power of procreation, even though **he** was too old—and Sarah herself was barren— because he considered him faithful who had promised." (NRSV)

Heb. 11:11 "It was faith that made **Abraham** able to become a father, even though **he** was too old and Sarah herself could not have children. He trusted God to keep his promise." (GN-TED)

Heb. 11:11 "Faith enabled **Abraham** to become a father, even though **he** was old and Sarah had never been able to have children. **Abraham** trusted that God would keep his promise." (GW)

Other translations have Sarah replaced by Abraham as well. Now I have to make a confession. At first I was critical of these translations. I thought it showed contempt for God's Word. The audacity of changing "Sarah" to "Abraham" and "she" to "he" seemed enormous. But, as I thought more about it, I realized that these translators were doing the most honorable thing they could. They knew *KATABOLE SPERMA* did not mean "conceive." They were in the same situation as Arthur Custance. I think they knew it meant either deposit seed or destroy seed, but it made no sense that Sarah would receive the power to do either one. So, they did the best they could with their knowledge of conception. They translated *KATABOLE SPERMA* in a way that wouldn't make the Holy Spirit look stupid. They made up a translation that fit their knowledge of conception. Ahhhhhh, but their knowledge of conception was limited.

I think the Holy Spirit did something no one expected. His knowledge of conception is perfect, and I think He revealed a scientific clue, not just a linguistic clue. I think the Holy Spirit picked the word He wanted because it defined exactly what happened inside of Sarah. Fast forward from the 1st century, when no one knew the details of conception, to the 20th century and the invention of the electron microscope. We discover something very interesting about *KATABOLE SPERMA*. Catabolism of sperm is exactly what must take place in order for new life to develop. The egg is surrounded by a thick covering called the *zona pellucida*. The *zona pellucida* does some important things:

1) It protects the egg.
2) It produces biochemical signals that attract sperm to it.
3) It provides specific binding sites for sperm attachment.

4) It prevents further sperm attachment once a sperm attaches. (It undergoes a rapid electrical change, via the cell's sodium pumps, that closes the other binding sites.)
5) It releases chemical "triggers" that initiate the process of sperm **catabolism.**

The sperm's DNA, which is inside its nucleus, must get into the egg's cytoplasm. The chromosomes inside the sperm have to become linked with the chromosomes inside the egg. Even when attached to the *zona pellucida*, the sperm can't do that. The sperm's DNA is biologically "miles" away from the egg's DNA. So, the egg produces biological signals in response to the attachment of the sperm. These signals trigger the release of other enzymes that start catabolizing the sperm cell. The sperm has to be broken down into smaller parts because only the necessary components are allowed into the egg. In addition, the outer membranes of the sperm and the egg have to be catabolized. A portal must be created for the transfer of the sperm's nucleus. This portal is made by catabolizing the membranes of both the egg and the sperm. This portal into the egg has to be just the right size too. The entire sperm isn't allowed into the egg because such an event could trigger a fatal reaction. Only the nucleus, the centrioles, and the flagellum are allowed to enter. The nucleus is needed because it contains the chromosomes. The centrioles are needed because they help position the chromosomes into their proper positions. The flagellum is needed to move the nucleus into place. The rest of the sperm is no longer needed. **The only way to separate the needed parts of the sperm from the unneeded parts of the sperm is by the process of catabolism.** If the woman's reproductive system doesn't have the ability to catabolize sperm, she will not be able to become pregnant. If the sperm can't be catabolized, its nucleus can't get inside the egg's cytoplasm. Only after the sperm cell is broken down into its smaller parts is it possible for the sperm's nucleus to be deposited into the egg. Once the sperm's nucleus is laid down inside the egg's cytoplasm, its chromosomes line up in their proper place. Once the father's chromosomes are in their proper place, they link with the chromosomes of the mother. Once the chromosomes link together, their DNA combines and conception takes place. Only if this breaking down/depositing process occurs can conception come about. Only after *KATABOLE SPERMA* do we witness the miraculous creation of a new… income tax deduction.

I believe the Holy Spirit picked *KATABOLE SPERMA* because that term defines with scientific precision, a scientific process that had to take place in order for Sarah to become pregnant. *KATABOLE SPERMA* is not some vague, mystical, symbolic figure of speech. It describes exactly what we learned about conception once we viewed it under the electron microscope. I believe the Holy Spirit was giving us another pre-scientific revelation. I believe He revealed that the "breaking down/laying down" of sperm is necessary for conception. I also believe He didn't allow that revelation to be understood until the electron microscope was invented. You see, I think the Holy Spirit likes electron microscopes. I think He thinks they're neat. I think He thinks they're neat because they allow us to observe things about His creation that we couldn't see otherwise. I think He thinks the same thing about the Hubble Space Telescope. I think the Holy Spirit rejoices when scientists reveal new truths about creation. I think God likes scientific discoveries. I think God likes scientists. Every time they discover a new truth about God's handiwork, it allows us to see something about Him. As a Christian, I think we should rejoice when scientists, even those who are atheists, discover new truths about nature. I don't think we should go around bashing scientists. It certainly doesn't help them see their discoveries as insights into the character of God. If you are a scientist and an atheist, then let me ask you to look a little closer at what you see in the universe around you. I think you will begin to see His beauty and His power and His Love if you look more closely at His work. I think you will see the same thing if you look more closely at His Word. It saddens me to see you miss God's miracles when you observe the universe. It saddens me even more that you don't feel the joy of your Creator when you discover a new truth about the universe. It saddens me most to know that you won't be able to spend eternity with your Creator learning more and more about Him and His universe. None of your questions about the universe will be answered in Hell.

Some will accuse me of playing the "What-If Game." Some will accuse me of trying to stretch the meanings of words to get the Bible to say what I want it to say. Well, I think they would be right if there was no catabolic process truly involved during physical conception. Many creationists have no problem with believing things that are absolutely contrary to scientific observations. They have no difficulty with biblical interpretations that say things like the speed of light has changed. They aren't bothered by translations that require fruit-bearing trees to have existed billions of years before the sun was created. With no biblical or scientific evidence to support their views, some scholars see nothing wrong with interpreting Genesis as saying there was a glowing ball of matter somewhere in space that gave light to the earth for three days before the sun was created. Some scholars believe it is perfectly okay to replace Sarah with Abraham when the Bible says something they don't want to believe. All these things are scholarly and noble and justified. But let someone imply that the Holy Spirit may have used a scientifically observable process to give us a clue that defends the Gap Theory, and suddenly that idea is denounced as unscholarly. Why do I think *KATABOLE*

315

SPERMA is a pre-scientific revelation of a scientific fact? It's because I don't believe in coincidences. It seems too coincidental for the Holy Spirit's choice of words involving Sarah's conception to accidentally describe a scientific phenomenon involving Sarah's conception. Imagine you discovered that your dog was dead, and you brought it to me for a necropsy, (a post-mortem examination) and I gave you this report: "Upon examination of the dog's thoracic cavity, I discovered that a .22 caliber bullet had penetrated the chest on the right side between the 5th and 6th intercostal space and severed the pulmonary artery. Cause of death was due to massive internal hemorrhage and hypovolemic shock." With that report in hand, would you think I was using words that described an actual medical event or words that were symbolic of some mystical, magical, indefinable phenomenon? I hope that my use of precise medical terms would clue you into the fact that I wasn't being mystical. I know what those terms mean and if I had intended to give you some vague report, I wouldn't have used them. I think the Holy Spirit wouldn't have used a precise biological term if He had meant to convey anything else but a precise biological process. I think this is why He said *KATABOLE SPERMA* instead of *SULLAMBANO SPERMA*. I think this was His clue to the meaning of *KATABOLE* in the Bible. I think He is telling us that *KATABOLE* means catabolism. *KATABOLE SPERMA* means the catabolism of sperm. It defines a true biological event. If I were a first century Greek scientist, and I knew what happened during conception, I would use the words *KATABOLE SPERMA* to describe the catabolism of the sperm cell that takes place inside the woman. *KATABOLE* means catabolism. *KATABOLE KOSMOS* means the catabolism of the world. It defines a true historical event. Once again, I think He chose His words to match His work. *KATABOLE KOSMOS* means there was a time in the past when the world was destroyed so that it could be rebuilt into a new and different world. *KATABOLE* describes both the tearing down of something old and the laying down of something new. That's what happened to the Pre-Adamic earth. There is no better term than *KATABOLE* to describe "Ruin-Restoration." I think *KATABOLE KOSMOS* means, "The Ruin-Restoration of the world." The scholars can attack my translation if they want. They can respond however they please. My only response is to respond to their response with the response the Holy Spirit makes: The Holy Spirit defines *KATABOLE KOSMOS* as the catabolism of the world... checkmate!

Chapter Eleven:

Science and the Restoration Theory

—⁂—

So how old is the universe? The Restoration Theory makes no statements concerning the absolute age of the universe. The Restoration Theory is based on biblical statements and the Bible doesn't give an age for the universe. A biblical age for the universe can only be stated as a biblical interpretation, not a biblical declaration. It would have been nice if Moses had told us the universe was created two thousand years before he was born, or ten million years before, or ten billion years before, or something. If Paul had said the universe was 13,254,487,219 years old, then we wouldn't need to debate this. The Bible doesn't reveal when the universe and the earth were originally created, or when they were restored. As it is, the Restoration Theory cannot be used to prove any dates. No biblical creation theory can do that!

The Bible doesn't tell us the age of the universe. This means we have to turn to science for answers. **The age of the universe is a scientific matter, not a biblical one.** No matter how great a Hebrew and Greek scholar you are, no matter how many theology degrees you have, you aren't going to find an age for the earth by studying the Bible. If you want to know the age of the universe, then sooner or later, you must enter the realm of science. When you do that, the very first thing you'll notice is the discrepancy between the science of the Young-Earth creationists and the science of the Old-Earth creationists. Again, I want to point out that I'm dealing only with those opinions held by Bible-believing, Christian Creationists. The problem is not one of biblical interpretation alone. It is a problem of scientific interpretation as well. (Of course, I'd be naive to believe that people's interpretation of science wasn't influenced by their interpretation of the Bible, and vice versa.)

Young-Earthers use scientific dating techniques to prove the earth has celebrated only a few thousand birthdays. Old-Earthers use scientific dating techniques to prove we should be putting billions of candles on earth's birthday cake. Which is correct? The answer is both and neither! It depends on your assumptions. All dating techniques require faith in basic assumptions. Before you put your faith in a dating technique, you need to understand its assumptions. Is it correct to assume that rocks originally contained no radioactive decay products at creation? Are all radioactive decay rates constant? Has cosmic ray bombardment always been the same? Does the $^{14}C/^{12}C$ ratio in the atmosphere vary over time? Has the earth's spin always been slowing at the same rate? Has space dust been accumulating on the earth and moon at constant rates? Whose rates do you believe? Is it valid to assume that earth's magnetic field has always been decaying the way it is today? Can you always use a present day change to extrapolate backward through time to achieve an absolute age? Did God create the universe with an apparent age, and if so, what apparent age? In all honesty, I've looked at the various dating techniques, thought about their assumptions, and have come away scratching my head. Now, it is true that both sides often use dating techniques that are so assumption-dependent and biased that they're easy to discount. It's also true, unfortunately, that at times both sides use known false data to persuade the unsuspecting. Still, both sides have more than a few dating techniques that seem valid, at least to me. Each side seems to be able to prove its point to a certain degree.

The Young-Earthers will shake their heads in disbelief. How could I even think the Old-Earth dating techniques might be valid? I'm sure I'll be accused of trying to make the Bible fit the Theory of Evolution. They'll point to a dozen books, written by Young-Earth creationists, that prove the Old-Earth dating techniques are invalid. The Day-Agers will do the same thing. They'll accuse me of being unscientific. They'll point to another dozen books, written by Day-Age creationists, that prove the Young-Earth dating techniques are invalid.

How are we going to resolve this? Who's right? Whose assumptions are we going to believe? Both sides can't be right... or can they? Like I said, I think both sides are partially right. Generally, I think both sides are making correct observations, but incorrect interpretations. This problem can be approached in three different ways.

1. We can assume the heavens and the earth were created only a few thousand years ago. With that assumption as a foundation we can begin searching for the evidence that supports this view.

2. We can assume the universe is billions of years old, and from there set out in search for supporting evidence.

Both ways of approaching the problem are scientifically valid. This is the scientific method. Assume the hypothesis, test the hypothesis, and then accept or reject the hypothesis in light of the evidence. (Of course in the real world, scientists, including Creation-Scientists, don't like to reject their own cleverly devised ideas.) Both ways are equally biblical as well. The Bible gives no specific dates. Dates can be assumed, but no verse in the Bible states an age for the universe. Unfortunately, some creationists act as if the Bible does. They confuse their assumption of a date with an actual Biblical statement of a date. They make their assumed date the criterion for judging both the scientific and biblical evidence. If the evidence doesn't fit their assumption, then the evidence is wrong.

3. A third way to approach this is not to assume any age for the heavens and the earth and let the scientific evidence say what the scientific evidence says.

As you can guess, I prefer this third approach. I prefer it because it doesn't have a built-in bias toward rejecting evidence that doesn't fit. If I were a Young-Earth Creationist, I'd have a tendency to assume all the Old-Earth dates were wrong. If I were a Day-Age Creationist, I'd have a tendency to assume all the Young-Earth dates were wrong. I'm a Restored-Earth Creationist, so that puts no limits of time on me. The heavens and the earth can be just as young or just as old as science finally proves them to be. Now, let me ask you something. What could we conclude if science really could prove some of the Old-Earth dates AND some of the Young-Earth dates? How would such proof affect the Restoration Theory? Finding proof for both ages of the earth would actually verify it. If the Restoration Theory is true, then one would expect to find just such evidence. Because the earth had two different beginnings, there would be both old and young dates for things. Conflicting dates wouldn't have to generate conflict. Suddenly, everything seems to fall in line with the Restoration Theory. If the Bible said the earth had only one beginning billions of years ago, then all the Young-Earth dating techniques would be wrong. If the Bible said the earth had only one beginning six thousand years ago, then all the Old-Earth dating techniques would be wrong. But if the Bible said the earth had two beginnings, then it might be possible for neither side to be completely wrong. The Day-Age Theory is not the true antithesis of the Young-Earth Theory. They don't have to conflict. The problem may not be with what the Bible and science teach. The problem may be with what people want the Bible and science to teach. The problem is with our motives. What motivates someone to believe a particular view? Why do people believe what they believe?

When you ask a Young-Earth Creationist why he or she believes that Genesis should be translated in a Young-Earth fashion, the typical answer is that it's the most clear and straightforward interpretation. Young-Earth creationists argue that God wouldn't write something that a dear, sweet, innocent, uneducated little-old-grandmother sitting in her rocking chair, reading her Bible, couldn't understand. Day-Age creationists do the same thing except it isn't Granny sitting in a rocking chair; it's a laboratory full of test tubes, telescopes, and microscopes that wind up being the interpreters of Genesis. God surely wouldn't write something contradictory to observable science.

Wrong on both cases! Consider the Old Testament prophecies of Christ and how many people (dear, sweet grandmothers included) couldn't understand them. God has also written things that contradicted observable science. In 2 Peter 3:10, Peter revealed that atoms could be split. That contradicted observable science until the 1940's when atomic fission was first observed. No, we must not insist a particular interpretation is false simply because it disagrees with our conclusions. Regrettably, this is what many Christians from all camps do. If the Scriptures destroy our interpretations and assumptions, then our interpretations and assumptions are wrong. If science destroys our interpretations and assumptions, then our interpretations and assumptions are wrong.

I know both Day-Age and Young-Earth creationists will object, but is the Restoration Theory really so objectionable?

For me, a Restored-Earth is not only not objectionable, it is the only thing that makes sense. I don't see anything wrong with assuming the Restored-Earth scenario, and then testing it against the dating techniques of the Young-Earth and Day-Age creationists. If you'll do this, you'll see that the Restoration Theory seems to fit both sets of data. I think many of the discrepancies in age arise because the two groups are observing two different events. The dating techniques may be correct but the conclusions are incorrect.

Living Fossils

The Restoration Theory explains many apparently contradictory observations. If God recreated life forms that had become extinct during the Pre-Adamic earth, then one could expect to find evidence of those life forms in both Old-Earth and Young-Earth settings. With this in mind, we see how the Restoration Theory explains "living fossils" such as the Coelacanth fish. Coelacanth apparently lived for a long time, but then died out seventy million years ago. Geology records a seventy million year absence of Coelacanth, yet living Coelacanth swim the ocean today. This is no problem for the Restoration Theory. What probably happened was that Lucifer brought about the extinction of Coelacanth seventy million years ago. Then suddenly, a few thousand years ago, God recreated Coelacanth so fishermen could catch them off the Acapulco Trench in 1939. Scientific observation supports the Restoration Theory. The evidence is in favor of a recent creation of life forms that had died out ages ago. There were Coelacanth fish on this earth a long time ago. Then they died out. The fossil evidence shows that no Coelacanth fish lived on earth for millions of years. Yet, here they are today. How can that have happened? They had to be recreated! The same can be said of Tuatara lizards, Ginkgo trees, Metasequoia trees, *Neopilinia*, and *Lepidocaris*. In 1994 an "extinct" species of pine tree, the Wollemi pine, was discovered in New South Wales.[46] This tree supposedly had been extinct for over sixty-five million years. This tree flourished millions of years ago. Suddenly there were no more Wollemi pines. They simply vanished. It's hard to imagine they survived millions of years without leaving evidence of their existence. Surely there would be Wollemi trees, branches, needles, cones, bark, or pollen somewhere in the fossil record if they had been alive. Scientific observations showed that Wollemi trees were missing from the fossil record for over sixty-five million years. How could Wollemi trees reappear on earth if they weren't recreated by God? The Restoration Theory explains what happened. Lucifer's manipulations of the earth brought about the extinction of Wollemi pines sixty-five millions of years ago. For sixty-five million years there were no Wollemi pines. Then a few thousand years ago, God restored them along with the other vegetation He restored on day three. The Gap Theory fits this scientific observation. These "living fossils" died out millions of years ago. The geological column faithfully recorded their absence. They were dead and gone! The geologists and paleontologists weren't wrong about their extinction. Then, Wollemi pines were recreated during a six day period sometime in the last several thousand years. The Young-Earthers aren't wrong about their recent creation. Restoration Creationists don't have to reject any proven data, biblical or scientific. In fact, if the Restoration Theory is true, then the Bible has given us a pre-scientific clue into the science of geology. If Satan brought death and devastation to the Pre-Adamic earth, then one could expect to discover geological evidence of death and devastation before man's existence. This is precisely what geology reveals. Geology therefore, is not only a valid scientific discipline, it is a vital key to interpreting the biblical creation account informationally.

Does The Gap Theory Allow for Evolution?

If the earth could be billions of years old, does it mean that evolution could have occurred during the gap? No, there was no evolution during the gap (everything was dead) and no evolution after the gap. All life was created/recreated during days 3-6, a period of 96 hours. How about evolution before the gap? Could God have "created" life in the original earth by the process of evolution? I would love to point to the Bible verse that says, "Knowest thou not that God didst not use evolution to create life in the Pre-Adamic earth?" Unfortunately, I can't find that verse. The Bible is silent on how God created life on the original earth. If I claimed that the Bible prohibited evolution in the Pre-Adamic world, I would be guilty of confusing an interpretation with a Biblical declaration. The only thing the Bible prohibits is evolution during and after the gap. Am I capitulating to the evolutionists? No! We must sift our biblical theories through the filter of science. I am convinced by SCIENCE that evolution never happened. I believe the scientific facts prove that evolution never occurred. (I explain this in *Objections to the Doctrine of Evolution*, but you can find the same evidence in many creationists' books.) It would be foolish to think the Bible taught something contrary to true science.

Was There A Big Bang?

If the universe is old, does it mean there was a Big Bang? Personally, I believe there was. The scientific evidence seems pretty solid, and the Bible may give us a hint in the affirmative. There are some verses that talk about how God "stretched" and is "stretching out" the heavens. (Psa. 104:2, Isa. 40:22, Isa. 42:5, Isa. 44:24, Isa. 45:12, Isa. 51:13, Jer. 10:12, and Zec. 12:1) Some people, including me, interpret these to indicate that the universe has been expanding from an original point of creation. It seems to make sense, but I will be the first to admit that it is an interpretation. I also admit that there are a lot of problems with the Big Bang Theory, but they are scientific problems, not biblical. Most of the problems with the Big Bang Theory are based on what astronomers and astrophysicists are discovering with further scientific observations and studies. The Restoration Theory doesn't say if the universe began with a Big Bang or not. Whether or not there was a Big Bang is something only science can discover. The Bible does not reveal the mechanics of how God created the universe in Genesis 1:1. A Big Bang, followed by a formation of the universe over billions of years could have been exactly how God created the heavens and earth in the beginning. Such an event would explain why God uses *BARA* (create) and *ASAH* (make) and *YATSAR* (form) to describe the creation of the universe. The Big Bang, however, was not how He restored the earth and not how He created the life now on it. The details of the Big Bang are something that can be learned only by continued scientific studies, not by theological arguments.

How Long Was the Earth *TOHUW WAW BOHUW*?

If there was a gap of billions of years between Genesis 1:1 and Genesis 1:3, was the earth *TOHUW WAW BOHUW* for billions of years? No. The Bible doesn't say this, and there is nothing in the geological record indicating that the earth lay desolate and dead for millions or billions of years. I would be foolish to let my biblical theory go unfiltered by scientific fact. While there may have been millions of years of life and death on earth before it was judged, the Bible does not reveal how long the judgment period itself was. More than likely the earth was *TOHUW WAW BOHUW* for a very short period of time. Various translations of Genesis 1:2 seem to indicate that the ocean was raging or roaring. If this was so, then it was still liquid. If the earth had been without the sun for very long, the ocean, or at least the surface of the ocean, would have been frozen. I believe the *TOHUW WAW BOHUW* period was so short that it would not have left a geological footprint. Personally, I think God judged the earth and let it lay dead for three days and three nights, and then began restoring it on a Sunday morning, the first day of the week. Such a scenario would have tremendous parallels with the resurrection of Christ. Nevertheless, the Bible doesn't say how long the earth was *TOHUW WAW BOHUW*. The only thing the Bible says is that God restored the earth during six literal days.

Alley Oops

It's virtually impossible to discuss life's origin without having the subject of "ape-men" popping up. Paleontologists have discovered fossils of ancient creatures that seem to be neither human nor ape. Evolutionists assume they're evolutionary links between apes and man. Young-Earth and Day-Age creationists disagree. So do Restored-Earth Creationists. (If they were Pre-Adamic, then they can't be man's ancestors because everything died when God judged the earth.) So what or who were these "ape-like/man-like" creatures? I see only four possibilities.

1) The first is that they were descendants of Adam and not Pre-Adamic. This requires you to reject many dating techniques and their basic assumptions. Some people believe that in spite of the morphological changes and aberrations, these creatures were truly human. Such bone deformities might be due to trauma, malnutrition, disease, arthritis, the environment, selective inbreeding, and a number of other factors. This has long been the favorite answer among creationists; especially in the case of Neanderthal man. Many

creationists believe that Neanderthal men were truly human because of their use of tools and their custom of burying their dead. Many said that if you gave a Neanderthal man a bath, a haircut and shave, and dressed him in a nice business suit, he would blend in if he walked across a typical college campus. That is ridiculous! Anyone bathed, shaved, well-groomed, and in a nice business suit wouldn't blend in on a typical college campus. Evolutionists said Neanderthal man wasn't human. Creationists said he was, and until recently, I agreed. I've now changed my mind. Recent DNA studies have shown that Neanderthal didn't have human DNA.[47,48] Neanderthal man doesn't appear to have been human. Now, that doesn't make him a link between ape and man, as evolutionists suggest, but it does mean he was not a descendant of Adam. That, and the fact that the scientific evidence seems to prove these creatures lived long before the time of true humans.

2) These "ape-men" could have been nothing more than unknown species of apes. There is a great deal of truth in this theory. Virtually all of the "ape-men" of the past have since been proven to be nothing more than apes, or even less. It's easy to place the Australopithecines and others in this group, but Neanderthal man poses a problem. If he was an ape, then he was a very sophisticated ape. The tools he made included knives, scrapers, chisels, and cleavers. He made clothing from animal skins and constructed tent-like shelters. He even made a calendar by notching grooves into bones so he could keep track of lunar cycles. (We know they weren't real men because they didn't have girly pictures on those calendars.) There is at least one case where a Neanderthal amputated a diseased arm from one of his fellow Neanderthals. I find it difficult to believe that Neanderthal was an ordinary hairy ape.

3) These "ape-men" could have been a variety of man-like animals created by God. God created them; they lived;

then they died. Even though they may have had physical characteristics similar to apes and men, they were no relation to either. Neanderthal seems to fit this category best. They were simply highly intelligent creatures that God placed on this earth, but subsequently became extinct. Of course, some people think they may not be extinct. You only need to mention the words "Sasquatch" and "Yeti" and hundreds of Bigfoot Believers will inundate you with tons of evidence for the present-day existence of these creatures.

4) Another belief is that these were Pre-Adamic humans that God once created, or at least something very close to human. They could have been the ones who planted gardens and fruitful places, and built cities. The Bible gives no description of the creatures who built and lived in the Pre-Adamic cities, but as I have previously said, I find it hard to believe that God would warn us about the sins of Pre-Adamic angels, but not about the sins of Pre-Adamic people. This of course assumes they sinned. It's possible they were human in many aspects but lacked the character of being made in God's image. In that aspect, they would have been more like very intelligent, very sophisticated animals that had no concept of moral right and wrong. They even may have been the unfortunate victims of Lucifer's attempts to rule the earth. If he was unsuccessful in complete domination of them, he may have waged war against them and killed them. He could have enslaved them. This could explain the comment in Isaiah 14:17 that he "opened not the house of his prisoners." Jesus could have been alluding to this when He said Lucifer was a murderer from the beginning. The murder/extinction of these innocent "people" could have been the final sin that caused God to destroy Satan's kingdom. It's all very interesting, but also very speculative.

Whatever was going on in the Pre-Adamic world, we must include Lucifer. If they truly were Pre-Adamic, then these creatures could have been any or all of the last three groups that came under the domination of Lucifer in his desire to have servants and slaves to worship him. It's possible he engaged in selective breeding and genetic alteration programs in order to develop creatures suited to his desires. He may have taken whatever kinds of apes, hominids, or "men" God had created and manipulated them for his own purpose. It's not as impossible as seems. Look at the variety of dog breeds we have "created" during our short history. They range from tiny things scarcely weighing three pounds, on up to huge one-hundred and eighty pound giants. Dogs have a tremendous variety of bones sizes and conformations. Some of their skulls are long and narrow; others are flat in the front and wide. Some have large cranial cavities and others have small ones. Some have very thick pelvises with flattened hip joints while others have nicely rounded ball-and-socket joints. This is what enables some to walk and run better than others. Greyhounds have long, thin, graceful hips and limbs, while Bassets waddle along on short, stubby, "twisted" appendages. They are all dogs, and all of their physical characteristics have been developed by selectively breeding according to our desires.

The variety of physical features seen in these fossils could be explained in a similar fashion. It doesn't seem far fetched to think that Lucifer may have monkeyed around with the most intelligent creatures God created in the Pre-Adamic world. God may have given these creatures enough intelligence and skill to make and use tools, plant gardens, build shelters, and develop a social order. (Even ants have a social order.) Lucifer may have taken advantage of their intelligence by teaching them to do the things he wanted them to do. Is that far fetched? Well, we've taught gorillas to speak sign language. We've taught dogs to guide the blind and to sniff out drugs and bombs. It doesn't take much imagination to believe that Satan may have spent millennia experimenting with these creatures until he developed some with hips, knees, pelvises, backs, and hands more suitable to serve him. Could Donald Johanson's "Lucy" (A 3.5 million year old *Australopithecus afarensis* skeleton found in Ethiopia in 1974) have been some poor, tortured creature that Lucifer bred into existence? What a tremendous discovery Johanson made if such is the case. If this is so, then paleontologists are digging up things far more spectacular than even they realize. **I think they are unearthing proof that the Bible is historically accurate when it says there was intelligent life before Adam.** What a testimony to the truth of the Bible if those fossils are the remains of intelligent Pre-Adamic life. Let's praise the paleontologists for their discoveries, not condemn them.

Chapter Twelve:

The Meaning of HAYAH

—⧑—

The evidence from qualified Hebrew scholars is that *HAYAH* can be translated "became" or "had become" in Genesis 1:2. I believe the evidence shows that *HAYAH* should be, not just could be, translated that way. If we are seeking an informationally correct translation rather than merely a grammatically correct translation, then we must be willing to look afresh at God's choice of this word in this passage. WE MUST NOT LET OUR PRECONCEIVED IDEAS OVERSHADOW WHAT GOD REVEALS. What information was God trying to communicate? Genesis was written in ancient Hebrew, to ancient Hebrews, for ancient Hebrews, by an ancient Hebrew. It seems to me that the only way to understand this Divine communiqué is by fully appreciating the mindset of the ancient Hebrews. I'm not sure that is possible. However, the more we think like ancient Hebrews, the better we will be able to translate Genesis 1:2 informationally. This task goes far beyond finding seven verses where *HAYAH* is used one way, thirteen verses where it is used another way, and another sixteen verses where it is used in a third way. We need to dig deeper than that. We need to understand what thought process went through the minds of the ancient Hebrews when *HAYAH* was employed. What did the word mean to them? Look again at what *Strong's Dictionary of the Hebrew Language*[49] said about *HAYAH*.

"H1961. *hayah*, haw-yaw'; a prim. root; to exist, i.e. be or become, come to pass (always emphatic, and not a mere copula or auxiliary)..."

James Strong said that *HAYAH* is not a mere copula. This is important. A copula is a word, usually a verb, that connects or links a subject to a predicate. I gave the example of how in the Hebrew thought process I would say, "Linda Dill my wife," rather than, "Linda Dill is my wife." This may sound strange to us modern English speakers, but ancient Hebrew people didn't think like we think. They didn't think in terms of *HAYAH* being a connecting word. This means we should be cautious in translating *HAYAH* in the sense of a connecting word. Remember what Martin Anstey expressed in *The Romance of Bible Chronology*.[50]

"When a Hebrew writer makes a simple affirmation, or merely predicates the existence of anything, the verb *hayah* is never expressed. Where it is expressed it must always be translated by our verb to become, never by the verb to be, if we desire to convey the exact shade of the meaning of the Original."

I want to point out something about the English word "was." "Was'" has numerous meanings. Look it up in a dictionary and you can easily find several definitions. Even if I translated Genesis 1:2 as, "and the earth WAS without form, and void," it could still describe a dynamic condition. Examine this sentence: "I broke my wife's favorite flower vase, and she was mad." What does "was" mean in this sentence? It obviously means that she got mad. She became angry. She wasn't in a constant, unchanging state of anger. Because I broke her favorite vase, she changed from a condition of not being mad to a condition of being mad. We understand the meaning of "was" because we understand the context of the sentence. We understand the true meaning of my sentence because we automatically determine the exact shade of meaning of "was." You would get the wrong idea about my wife if you failed to understand the various meanings of "was." If you translated my sentence into another language, you could produce a grammatically correct translation that failed to convey the truth. If you didn't understand the many nuances of "was" in English, you could make people think she was always mad. The only way to translate my sentence correctly is if you understood the English-speaking mind. If you didn't think like an English speaker, you wouldn't be able to translate it in a way that captured the correct information. The same is true of ancient Hebrew. We can't fully understand ancient Hebrew if we don't fully understand the ancient Hebrew mind. So, let's delve deeper into the ancient Hebrew mind. I think this comment from *The Complete Word Study Old Testament, King James Version*[51] will help clarify the ancient Hebrew thinking process concerning the idea of being or existence when the verb *HAYAH* was used.

"1961. Hāyāh; probably related to hāwāh (1933), "to breathe." This verb means to exist, to be, to become, to come to pass, to be done, to happen, to be finished. It is notable that this verb was not used in a copulative construction in Hebrew. Bowman maintains that the Hebrews thought only in dynamic categories, not static ones."

What does it mean that they thought only in dynamic categories, not static ones? It means they used *HAYAH* when they wanted to express the idea of dynamic existence. Generally, they didn't use it when the thought of static existence was to be expressed. Let me show how English thought differs from Hebrew thought. Consider these two statements:

1) I am Steven Dill.
2) I am president of the Jefferson County Veterinary Medical Association.

Does the "am" in statement one mean the exact same thing as the "am" in statement two? No, the first "am" conveys an idea of static existence. The second "am" conveys an idea of dynamic condition. I haven't always been the president of my local veterinary medical association. I won't always be its president. President or not, I remain Steven Dill. These represent two different categories of "being;" two different kinds of "am." In the minds of the ancient Hebrews, the concept of me "being" president of a local association would be expressed using *HAYAH*. It's a dynamic condition. *HAYAH* would not be used for the idea of me "being" Steven Dill. That's a static condition. Hebrew distinguishes between these two ideas, as do other languages. Spanish for instance, uses two verbs for "to be." The first is *SER* and the second is *ESTAR*. In Spanish I would ask you the question, *¿Que ES su apellido?* What is your surname? Your surname is more or less a permanent description or characteristic of you; therefore *ES* (a form of the verb *SER*) is used. In contrast, I would ask you, *¿Como ESTÁ usted?* How ARE you? Since your present condition of health is not so permanent, *ESTAR* is the verb of choice. Spanish goes a step farther by expressing some characteristics in terms of "having" rather than in terms of "being." In English, I would say, "I am hungry." In Spanish, I would say, *"Yo tengo hambre,"* which literally means, "I have hunger." I am not the same thing as hunger; hunger and I are not equivalent. I'm sure many other languages make these same distinctions in various ways. In English, however, we use the same word for both concepts of "being." I am not a Hebrew scholar, but since Hebrew scholars have said it, I can say it too: Generally, *HAYAH* is used for dynamic categories or conditions, and not as a static linking word. What kinds of conditions were generally considered dynamic? The answer is simple. Any condition that is not static, any condition that has changed, is changing, or will change, was thought of as being dynamic. Let's look at some examples.

HAYAH (in bold print) Used for Dynamic Conditions
of Being

Geographical Location

Gen. 38:5 "And she yet again conceived, and bare a son; and called his name Shelah: and he **was** at Chezib, when she bare him." (KJV)

HAYAH is used here because this Canaanite woman (Shuah's daughter) was at Chezib when she gave birth to Shelah, Judah's son. Geographical location was not thought of as a permanent condition or state. It was dynamic. The third *HAYAH* in Genesis 39:2 is another example of how location was thought of as a dynamic condition.

Gen. 39:2 "And the LORD **was** with Joseph, and he **was** a prosperous man; and he **was** in the house of his master the Egyptian." (KJV)

Joseph was in his master's house, but his location was not static. Position is dynamic. This is why *HAYAH* is used. If I wanted to tell you my current location in English, I would say, "I am upstairs." If I wanted to say it in Hebrew, I would want to use *HAYAH*. Without it, I could be misunderstood as saying that I am a hallway, three bedrooms, two bathrooms, and an office. *HAYAH* would indicate that "upstairs" is my dynamic position and not my static being.

Temporal Conditions

While we're looking at Genesis 39:2, note the first two *HAYAH*'s. Having the Lord's presence was a dynamic condition. The ancient Hebrews considered the Lord's presence as an indicator of His blessing and approval. They didn't think of God's presence with them as an absolute permanent, static condition. God was always with them in one sense, He is omnipresent, but He wasn't "with" them when they were outside His will. God's presence in their lives was dynamic. The second *HAYAH* in Genesis 39:2 was also a dynamic condition. Joseph's condition of prosperity was not permanent either. He would soon be cast into prison.

Age

Gen. 17:1 "And when Abram **was** ninety years old and nine, the LORD appeared to Abram, and said unto him, I *am* the Almighty God; walk before me, and **be** thou perfect." (KJV)

Apparently the ancient Hebrews generally viewed a person's age as a dynamic condition. I am 59, but my age is dynamic. Again, since I am a little more familiar with Spanish, I can understand why the Hebrews would think in this fashion. Spanish-speaking people don't express age the way English-speaking people do. If I gave my age in Spanish, I would say *"Tengo cincuenta y nueve años;"* which literally means, "I have fifty-nine years." To an ancient Hebrew my age would be dynamic. I have become 59. I haven't always been 59, and Lord willing, I won't always be 59. We could translate Genesis 17:1 as, "And when Abram became ninety-nine years old..." (Note that *HAYAH* is not used in, "I *am* the Almighty God," but it is used in, "**be** thou perfect." There is a distinction between God's being and our being that I'll explain shortly.)

Occupation or Activity

Age wasn't the only thing they viewed as dynamic. A person's occupation, job, or activity could also be considered dynamic. I am a veterinarian, but I haven't always been one. *HAYAH* would be used if I said, "I am a veterinarian." The tricky part in some of these conditions is trying to determine what it means to be something. There could be a difference in what is means by saying I am a veterinarian. In one sense I could say that I will always be a veterinarian even if I retire and no longer actively practice veterinary medicine. Yet, in another sense, if I am not actually being a veterinarian, I am not a veterinarian. Here is an example:

Gen. 40:13 "Yet within three days shall Pharaoh lift up thine head, and restore thee unto thy place: and thou shalt deliver Pharaoh's cup into his hand, after the former manner when thou **wast** his butler." (KJV)

Being Pharaoh's butler wasn't a permanent state for this poor fellow. He had been Pharaoh's butler, but now found himself in jail along with Joseph and the Pharaoh's former baker. He was no longer Pharaoh's butler. His occupation was a dynamic condition. *HAYAH* is used to express this idea. Fortunately for him, being in jail was also a dynamic condition. He was going to become Pharaoh's butler once again. (Unfortunately for the baker, he was going to be executed.) This concept isn't too difficult for us to understand. He WAS a butler one day and he WAS not a butler the next. His condition was dynamic. There are other examples that aren't as easily understood.

Some Physical Characteristics

Gen. 29:17 "Leah *was* tender eyed; but Rachel **was** beautiful and well favoured." (KJV)

Note that the first "was" is italicized. There is no "was" here; *HAYAH* is not used. Commentators describe Leah's condition as being one of poor eyesight. The ancient Hebrew mind apparently thought that poor eyesight was a permanent condition. *HAYAH* wasn't required to describe it. They might think differently today in our world of eyeglasses, contacts, and laser surgery. In contrast to Leah's eye condition, whatever it was, Rachel was beautiful and well favored. Outward beauty is a fading quality. I'm sure the ancient Hebrews understood that. Our culture certainly believes it. We spend billions of dollars every year trying to retain the appearance of youth and beauty. Rachel was beautiful, but outward beauty is not a permanent condition. Now, there could be another thought here. This verse could be implying that Rachel grew to become beautiful. The Hebrew words describe both her form and her face.

Gen. 29:17 "And Leah's eyes were weak, but Rachel **was** beautiful of form and face." (NASB)

Gen. 29:17 "Leah had weak eyes, but Rachel **was** lovely in form, and beautiful." (NIV)

Gen. 29:17 "And the eyes of Leah were tender; but Rachel **was** of beautiful form and beautiful countenance." (DNT)

Gen. 29:17 "and the eyes of Leah *are* tender, and Rachel **hath been** fair of form and fair of appearance." (YLT)

The emphasis of the dynamic *HAYAH* could be what Rachel came to be or what she became as she grew to womanhood. It seems to me that this is a reference to her form. Without trying to be crude, it seems to refer to her bodily form, her figure. She had a great figure and she was beautiful. Little girls are cute, they may even be beautiful, but having a beautiful form is something that comes to be. Anyway, whatever her conditions were, they were dynamic

Moral, Spiritual, or Behavioral Characteristics

Gen. 6:9 "These *are* the generations of Noah: Noah **was** a just man *and* perfect in his generations, *and* Noah walked with God." (KJV)

Noah was a just man, but since he was born of Adam's line, we know he was also a sinner. He was born a sinner. He wasn't always a just man, but he did become one. We also know that he wouldn't always remain in that state. He would sin again. *HAYAH* is used to express this kind of dynamic "was." God was pleased with Noah's goodness, **but see how God contrasts man's being with His own being.** Look at these different forms of being.

Lev. 11:45 "For I *am* the LORD that bringeth you up out of the land of Egypt, to be your God: ye shall therefore **be** holy, for I *am* holy." (KJV)

HAYAH is necessary to describe man being holy because being holy isn't something we are. It is something we can become. This passage could be translated, "ye shall therefore **become** holy." Now, look at the second "*am*." It is in italics: "for I *am* holy." No *HAYAH* is used because God's state of Holiness is static, unchanging, and permanent. God doesn't "become" Holy. God is not dynamically Holy. God IS Holy.

He always has been and He always will be. We are not holy in the same sense that God is Holy. Do you begin to see what *HAYAH* meant to the ancient Hebrews? This doesn't mean that *HAYAH* should always be translated "became." Other translations are equally valid, but we must remember that when they are used, generally they are not used to imply a condition of static being. Genesis 1:2 does not describe a static condition of the earth. It describes a dynamic condition, a condition that had changed. The earth had changed from a condition of not being "without form, and void," to a condition of being "without form, and void." This is what *HAYAH* conveys. This is what the ancient Hebrews would have thought.

Let's examine this in more detail. I have searched the Book of Genesis to find all the instances of *HAYAH*. As I said earlier, this was done manually and not by computer, so it is quite possible that I missed some. I know that somewhere, some Hebrew scholar has already done this and did a much better job, but I don't have that information before me. Besides, I wanted to make sure for myself that what I have read about *HAYAH* was accurate. So, even if I didn't find all the uses of *HAYAH*, I think you will still see how *HAYAH* should be translated "became" or "had become" in Genesis 1:2. What follows is a list of the English translations (KJV) of *HAYAH* from the Book of Genesis. The list is in alphabetical order, and I included in brackets [ ] the number of times *HAYAH* is translated the way it is. Note that I have marked the first listing with an asterisk. These eight instances are cases where *HAYAH* is actually in the Hebrew text, but no form of the verb is translated directly in the *King James Bible*. In these cases the *HAYAH* is indirectly expressed in the context of the translation. I'm not going to worry about how *HAYAH* can NOT be translated; I have enough problems with how it CAN be translated.

HAYAH in Genesis (KJV)

* [8] 1:5, 1:8, 1:13, 1:19, 1:23, 1:31, 15:12, 32:8
are [3] 42:11, 42:31, 42:36
be [3] 17:1, 24:60, 38:23
became [9] 2:7, 2:10, 19:26, 20:12, 21:20, 24:67 47:20, 47:26, 49:15
become [8] 3:22, 9:15, 18:18, 32:10, 34:16, 37:20, 48:19, 48:19
been [1] 47:9
built [1] 4:17
came [1] 15:1
came to pass [64] 4:3, 4:8, 4,14, 6:1, 7:10, 8:6, 8:13, 11:2, 12:11, 12:14, 14:1, 15:17, 19:17, 19:29, 19:34, 20:13, 21:22, 22:1, 22:20, 24:15, 24:22, 24:30, 24:52, 25:11, 26:8, 26:32, 27:1, 27:30, 29:10, 29:13, 29:23, 29:25, 30:25, 30:41, 31:10, 34:25, 35:17, 35:18, 35:22, 37:23, 38:1, 38:9, 38:24, 38:27, 38:28, 38:29, 39:5, 39:7, 39:10, 39:11, 39:13, 39:15, 39:18, 39:19, 40:1, 40:20, 41:1, 41:8, 41:13, 42:35, 43:2, 43:21, 44:24, 48:1
come to pass [9] 4:14, 9:14, 12:12, 24:14, 24:43, 27:40, 44:31, 46:33, 47:24
continually [1] 8:5
continued [1] 40:4
had [6] 11:3, 11:3, 12:16, 13:5, 26:14, 30:43
had been [2] 13:3, 31:42
hadst [1] 30:30
hath been [3] 31:5, 46:32, 46:34
have [1] 32:5
keep [1] 33:9
let [1] 1:6
let be [10] 1:3, 1:6, 1:14, 1:14, 1:15, 13:8, 24:51, 26:28, 31:44, 37:27
may be [1] 21:30
mayest be [1] 28:3

might be [1] 30:34

seemed [2] 19:14, 29:20

shall be [45] 1:29, 2:24, 3:5, 4:14, 6:3, 6:19, 6:21, 9:2, 9:3, 9:11, 9:13, 9:16, 9:25, 9:26, 9:27, 15:5, 15:13, 17:5, 17:11, 17:13, 17:16, 17:16, 27:33, 27:39, 28:14, 28:21, 28:22, 30:32, 31:8, 31:8, 34:10, 35:10, 35:11, 41:27, 41:36, 41:36, 44:10, 44:10, 44:17, 47:24, 48:5, 48:6, 48:21, 49:17, 49:26,

shall have [1] 18:12

shalt be [6] 4:12, 12:2, 17:4, 24:41, 41:40, 45:10

shall seem [1] 27:12

should be [3] 2:18, 18:25, 38:9

to be [5] 10:8, 17:7, 18:11, 34:22, 39:10

was [68] 1:2, 1:3, 1:7, 1:9, 1:11, 1:15, 1:24, 1:30, 2:5, 3:1, 3:20, 4:2, 4:2, 4:20, 4:21, 5:32, 6:9, 7:6, 7:12, 7:17, 10:9, 10:10, 10:19, 10:30, 11:1, 11:30, 12:10, 13:6, 13:7, 15:17, 17:1, 21:20, 23:1, 25:20, 25:27, 26:1, 26:1, 26:28, 26:34, 27:30, 29:17, 30:29, 31:40, 35:3, 35:5, 35:16, 36:12, 37:2, 38:5, 38:7, 38:21, 38:22, 39:2, 39:2, 39:2, 39:5, 39:6, 39:20, 39:21, 39:22, 41:13, 41:53, 41:54, 41:54, 41:56, 42:5, 47:28, 50:9

wast [1] 40:13

were [37] 1:5, 1:8, 1:13, 1:19, 1:23, 1:31, 2:25, 4:8, 5:4, 5:5, 5:8, 5:11, 5:14, 5:17, 5:20, 5:23, 5:27, 5:31, 6:4, 7:10, 9:18, 9:29, 11:32, 25:3, 26:35, 27:23, 30:42, 34:5, 34:25, 35:22, 35:28, 36:7, 36:11, 36:13, 36:14, 36:22, 41:48

will be [9] 16:12, 17:8, 26:3, 28:20, 31:3, 34:15, 44:9, 47:19, 47:25

HAYAH is translated "became" or "become" 17 times while it is translated "was" or "wast" 69 times. I think at this point some people will say, "See, there are 69 votes for 'was' and only 17 votes for 'became.'" From this they might conclude that "was" is the winner. They might conclude that Genesis 1:2 must be translated "was." But if you include the phrases "came to pass," "come to pass," and, "came," which mean the same thing as "became" and "become," then you add 74 translations in favor of "became." This makes a total of 91 times that *HAYAH* is used to mean "coming to be," a dynamic being and not a static being. *HAYAH* is used 22 more times to describe a form of "coming to be" than it is used to describe a form of "was." How can anyone say *HAYAH* can't be translated as "became?"

If I say, "the earth was without form, and void," what does "was" mean? Is "was" always "was" as some say it was, or was "was" something else that God said it was? (I've been sitting at this keyboard too long!) The *HAYAH* of Genesis 1:2 is translated "was." Some claim that this is the only way it can be translated. Is that true? Is this "was" the same kind of "was" as all the other translations of "was?" The only way to answer this is to look at all the *HAYAH*s in Genesis that are translated "was." We can then see if they always express the exact same idea or concept. Please note: I am only going to look at the Book of Genesis. I know this puts an artificial limit on how much we can learn about *HAYAH*, but to do a complete Old Testament study of *HAYAH* is beyond the scope of this book and beyond my abilities. Hebrew scholars have written numerous books on the use of *HAYAH*, so if you want to know more, you can easily continue your research. My purpose in writing this section about *HAYAH* is to prove three things.

1. Many of the "was" translations actually express an idea of "become" or "come to pass."
2. Most of the "was" translations are a different form of the verb than in Genesis 1:2.
3. The exact same form of *HAYAH* is elsewhere translated "become" or "became."

HAYAH Translated as "Was" or "Wast" in the KJV

WAS

Gen. 1:2 "And the earth **was** without form, and void; and darkness *was* upon the face of the deep. And the Spirit of God moved upon the face of the waters."

Since the *HAYAH* in Genesis 1:2 is the one in question, let's leave it until last. Let's skip ahead to *HAYAH* in Genesis 1:3.

Gen. 1:3 "And God said, Let there be light: and there **was** light."

HAYAH is used twice. "BE" and "WAS" are both translations of *HAYAH*. What is literally said is, "God said, 'light, COME TO BE' and light CAME TO BE." He spoke light into being. Light's existence was a dynamic condition. It came into being. We could substitute, "and it became light," and it would actually make more sense. The same can be said of God's other creative commands. They all CAME TO BE at God's command.

Gen. 1:7 "And God made the firmament, and divided the waters which *were* under the firmament from the waters which *were* above the firmament: and it **was** so."

Gen. 1:9 "And God said, Let the waters under the heaven be gathered together unto one place, and let the dry *land* appear: and it **was** so."

Gen. 1:11 "And God said, Let the earth bring forth grass, the herb yielding seed, *and* the fruit tree yielding fruit after his kind, whose seed *is* in itself, upon the earth: and it **was** so."

Gen. 1:15 "And let them be for lights in the firmament of the heaven to give light upon the earth: and it **was** so."

Gen. 1:24 "And God said, Let the earth bring forth the living creature after his kind, cattle, and creeping thing, and beast of the earth after his kind: and it **was** so."

Gen. 1:30 "And to every beast of the earth, and to every fowl of the air, and to every thing that creepeth upon the earth, wherein *there is* life, *I have given* every green herb for meat: and it **was** so."

All these things became so. They weren't always so. "Was" is a good word but not "was" in the static sense. These things were dynamic in their being. The Hebrews thought in dynamic categories. This is why they used *HAYAH*.

Gen. 2:5 "And every plant of the field before it **was** in the earth, and every herb of the field before it grew: for the LORD God had not caused it to rain upon the earth, and *there was* not a man to till the ground."

In Genesis 2:5 we see another *HAYAH*. Here again, it indicates a dynamic condition. The word "before" shows that it is dynamic with respect to time. Plants were not, but now they soon would be. Very, very carefully now, contrast this with the last part of this sentence: "*there was* not a man to till the ground." **No *HAYAH* is used here**; this "*was*" is in italics. Man was not created yet either. Earth was without plants and earth was without man, but *HAYAH* is not used in the case of man's not being. This was a static condition as far as the earth was concerned. There had never been man; man was not. These two different uses of "was" in the same sentence and in the same context indicate that there was a difference between earth <u>dynamically</u> being without plants and earth <u>statically</u> being without man. Whatever the ancient Hebrews thought about this, they made a distinction between plants "not being" and man "not being." They thought of "not being" in two different ways. The plants' "not being" was dynamic; it had changed. Man's "not being" was static; man had never been. This, I believe, is an indication that plant life had been there before, but wasn't (a dynamic situation) there now, while man had never (a static condition) been there. Such word usage helps support the Restoration Theory.

Gen. 3:1 "Now the serpent **was** more subtle than any beast of the field which the LORD God had made. And he said unto the woman, Yea, hath God said, Ye shall not eat of every tree of the garden?"

Was the serpent more subtle, or did the serpent become more subtle? This presents the very same problem as the earth being or becoming without form, and void in Genesis 1:2. There is very little context to give us the clues we need. What does subtle mean? The Hebrew word has two connotations, one is good and the other is bad. It can mean "prudent" or it can mean "crafty." In this context of deceiving Eve, I can't see how we can interpret this word except in its bad sense. The serpent "was" or "became" subtle, crafty, or shrewd. Had it always been that way? Had it always been able to speak? What kind of animal was this serpent? I have no answers. I know that the real culprit was Lucifer, the Great Deceiver, who is often called "the Serpent." I do not think the actual animal, whatever it was, was subtle in and of itself. It **became** subtle only by the power of Lucifer. How can an animal be subtle, especially a reptile? While I am not a specialist in reptile medicine, I can say this with authority: Reptiles don't have the neural capacity to be prudent, shrewd, subtle, or crafty. That's not what they are. That's not their static, permanent, ongoing condition or state. Even if this creature of God originally had cognizance and advanced intelligence, I know God didn't create it shrewd or crafty in an evil sense. It wasn't originally that way; it became that way. I don't think Lucifer possessed, indwelt, or controlled this reptile from the beginning of creation. I don't think this animal's relationship with Lucifer was static. I think it was dynamic. I think this reptile became subtle only when Lucifer acted upon it. The serpent, therefore, became subtle. This is what *HAYAH* expresses. I think *HAYAH* expresses the same thing about the earth becoming without form, and void. It wasn't its original condition.

Gen. 3:20 "And Adam called his wife's name Eve; because she **was** the mother of all living."

Now here's an interesting *HAYAH*. It is translated "was," but what does it really mean? Eve wasn't the mother of all living at this point in time. In fact, she wasn't yet the mother of anyone. However, she became the mother of all living. When we modern English speakers see "was" we tend to think in terms of the past. "I was in Scotland." "I was a sailor." "I was sleepy." Many creationists opposed to the Gap Theory say that this is the only way that *HAYAH* in Genesis 1:2 can be translated. This verse shows that they are wrong. It would be poor English to say, "I was in Chicago next year." It wouldn't make sense to us to use "was" for a future condition, yet here we have *HAYAH* used for a future condition. It's obvious that *HAYAH* doesn't always mean what some people want it to mean. Eve became the mother of all living.

Gen. 4:2 "And she again bare his brother Abel. And Abel **was** a keeper of sheep, but Cain **was** a tiller of the ground."

As was true for Eve, so it was for Cain and Abel. At his birth, Abel was not a keeper of sheep. He became a keeper of sheep. Neither was Cain a tiller of the ground when he was born. Cain became a tiller of the ground. "Became" is a much better translation; it expresses the information more accurately. This same concept of "was" as a future event is seen in the next two passages as well.

Gen. 4:20 "And Adah bare Jabal: he **was** the father of such as dwell in tents, and *of such as have* cattle."

Gen. 4:21 "And his brother's name *was* Jubal: he **was** the father of all such as handle the harp and organ."

Jabal and Jubal were apparently inventors. Jabal invented tents and Jubal invented musical instruments. At least this is what I have been told that being the "father of" something meant. They became inventors; they became the "fathers" of nomads (tent-dwellers with livestock) and musicians. There had been no nomads or musicians before this, but now there were. Whatever it meant, it didn't mean "was" in the sense of it being a static, past event. "Was" in these two verse means "became."

Gen. 5:32 "And Noah **was** five hundred years old: and Noah begat Shem, Ham, and Japheth."

This *HAYAH* expresses the dynamic concept of a person's age. I've already mentioned this before. From now on, I'll skip over the instances of *HAYAH* that have already been explained. I'll still list them, but unless I see some significant difference, I'll not bother re-explaining my thoughts.

Gen. 6:9 "These *are* the generations of Noah: Noah **was** a just man *and* perfect in his generations, *and* Noah walked with God."

Gen. 7:6 "And Noah *was* six hundred years old when the flood of waters **was** upon the earth."

Gen. 7:12 "And the rain **was** upon the earth forty days and forty nights."

Gen. 7:17 "And the flood **was** forty days upon the earth; and the waters increased, and bare up the ark, and it was lift up above the earth."

The Flood was not always upon the earth. It came upon the earth. The rain was not always on the earth. The rain came upon the earth for forty days and forty nights. These three verses in Genesis 7 express dynamic conditions.

Gen. 10:9 "He **was** a mighty hunter before the LORD: wherefore it is said, Even as Nimrod the mighty hunter before the LORD."

As we have seen, expressing a person's occupation, activities, and actions are dynamic expressions in Hebrew. Nimrod wasn't always a mighty hunter; he became one.

Gen. 10:10 "And the beginning of his kingdom **was** Babel, and Erech, and Accad, and Calneh, in the land of Shinar."

This *HAYAH* is dynamic because it describes the beginning of something. Nimrod didn't always have a kingdom.

Gen. 10:19 "And the border of the Canaanites **was** from Sidon, as thou comest to Gerar, unto Gaza; as thou goest, unto Sodom, and Gomorrah, and Admah, and Zeboim, even unto Lasha."

Gen. 10:30 "And their dwelling **was** from Mesha, as thou goest unto Sephar a mount of the east."

These two verses indicate geographical locations. A person's geographical location or position was a dynamic condition in Hebrew thinking. Locations and boundaries change.

Gen. 11:1 "And the whole earth **was** of one language, and of one speech."

At first glance it may seem like we have finally found *HAYAH* used in a static sense. This isn't a situation where the population of the world became of one language. It was already one language. Multiple languages may not have existed until after God confounded their speech at the Tower of Babel. If this is true, then it would be wrong to translate this, "And the whole earth became of one language..." So wouldn't it be equally wrong to translate Genesis 1:2 as "became?" No, this *HAYAH* has an altogether different meaning here. The key to interpreting this verse correctly is to remember that *HAYAH* has multiple meanings. It doesn't always translate as some form of "to be." It can also be translated as a form of "to have." Go back and look at the list of different translations of *HAYAH* used in the *King James Bible*. Don't let these other meanings of *HAYAH* throw you off track. Words that are spelled the same and sound the same don't always mean the same. We see this in English. "Tire" means to become exhausted, but a "tire" is something you put on your car. "Fire" means something is burning up, but "fire" also is what a boss does to an incompetent employee. The *HAYAH* in this verse is different than the *HAYAH* in Genesis 1:2. It does not mean that the people and the language were identical. It does not mean they were a language. It means they had one language. This is the way *The New International Version, The New Revised Standard Version, The New King James Version,* and several other versions translate it. *The New American Standard Bible* doesn't use "was" either.

Gen. 11:1 "Now the whole world had one language and a common speech." (NIV)

Gen. 11:1 "Now the whole earth had one language and the same words." (NRSV)

Gen. 11:1 "Now the whole earth had one language and one speech." (NKJV)

Gen. 11:1 "Now the whole earth used the same language and the same words." (NASB)

I think I would be on safe ground saying this passage is best understood as "the whole earth HAD one language," rather than "the whole earth WAS one language." These Hebrew scholars say so. You may disagree. That's your choice, but I have yet another Hebrew scholar who interpreted it in this fashion. I would be careful disagreeing with Him.

Gen. 11:6 "And the LORD said, Behold, the people *is* one, and they **HAVE** all one language; and this they begin to do: and now nothing will be restrained from them, which they have imagined to do." (KJV)

I'm sure that He who created the Hebrew language understood the Hebrew language. God offers His own interpretation of Genesis 11:1 in Genesis 11:6. He says, in different words so as to make it clear, that they all "HAVE" one language. He doesn't use *HAYAH* in Genesis 11:6. (Note that even the "*is*" is in italics.) Genesis 11:6 gives us God's interpretation of Genesis 11:1. The *HAYAH* in Genesis 11:1 doesn't have the same meaning as the *HAYAH* in Genesis 1:2. It doesn't mean "to be," it means "to have." Therefore, no one can use Genesis 11:1 as evidence against the Gap Theory. Some have tried this, but the only way they can do it is to interpret the *HAYAH* in Genesis 11:1 in a way that not even God interprets it.

Gen. 11:30 "But Sarai **was** barren; she *had* no child."

It's true that Sarah didn't become barren. She already was barren. Being barren, however, is a medical condition and such conditions were often thought of in terms of being dynamic. Now, I'm not sure how the ancient Hebrews viewed all the different medical conditions. We have seen evidence that they viewed Leah's poor vision as a static, non-changing condition. Whatever their thinking was about other medical maladies, I know that barrenness was viewed as curse from God. To them it indicated, rightly or wrongly, God's disapproval of them or of something in their lives. This mindset pervaded their culture. It was a shame for a woman to be barren. Having children was, in their thought process, an indicator of God's presence in their lives. Being barren was an indicator that God was not with them. As we have seen before, the Hebrews didn't think of God's approval as a static condition. The same was true of God's disapproval. They didn't think it was impossible to appease God and regain His presence and approval. They prayed, they obeyed the Law, they performed the rituals and offered the sacrifices hoping for this very thing. I cannot be certain, but it seems to me that when a woman was barren, she prayed and hoped for it to be a dynamic state, not a static one. This appears to be the thought process in the women mentioned in the Bible who were barren. However they thought about it, in Sarah's case we know that it was a dynamic condition. She was barren at this point but she wouldn't always be barren. One day she would give birth to Isaac. Because of this, the *HAYAH* in Genesis 11:30 is also a dynamic "was."

I will skip over the next several verses because we have already discussed similar uses of *HAYAH*. I want you to look at them, however, so you can see for yourself how "came," "came to be," or "became" can often be a better translation.

Gen. 12:10 "And there **was** a famine in the land: and Abram went down into Egypt to sojourn there; for the famine *was* grievous in the land."

Gen. 13:6 "And the land was not able to bear them, that they might dwell together: for their substance **was** great, so that they could not dwell together."

Gen. 13:7 "And there **was** a strife between the herdmen of Abram's cattle and the herdmen of Lot's cattle: and the Canaanite and the Perizzite dwelled then in the land."

Gen. 15:17 "And it came to pass, that, when the sun went down, and it **was** dark, behold a smoking furnace, and a burning lamp that passed between those pieces."

Gen. 17:1 "And when Abram **was** ninety years old and nine, the LORD appeared to Abram, and said unto him, I *am* the Almighty God; walk before me, and be thou perfect."

Gen. 21:20 "And God **was** with the lad; and he grew, and dwelt in the wilderness, and became an archer."

Gen. 23:1 "And Sarah **was** an hundred and seven and twenty years old: *these were* the years of the life of Sarah."

Gen. 25:20 "And Isaac **was** forty years old when he took Rebekah to wife, the daughter of Bethuel the Syrian of Padanaram, the sister to Laban the Syrian."

Gen. 25:27 "And the boys grew: and Esau **was** a cunning hunter, a man of the field; and Jacob *was* a plain man, dwelling in tents."

Gen. 26:1 "And there **was** a famine in the land, beside the first famine that **was** in the days of Abraham. And Isaac went unto Abimelech king of the Philistines unto Gerar."

Gen. 26:28 "And they said, We saw certainly that the LORD **was** with thee: and we said, Let there be now an oath betwixt us, *even* betwixt us and thee, and let us make a covenant with thee;"

Gen. 26:34 "And Esau **was** forty years old when he took to wife Judith the daughter of Beeri the Hittite, and Bashemath the daughter of Elon the Hittite:"

Gen. 27:30 "And it came to pass, as soon as Isaac had made an end of blessing Jacob, and Jacob **was** yet scarce gone out from the presence of Isaac his father, that Esau his brother came in from his hunting."

Gen. 29:17 "Leah *was* tender eyed; but Rachel **was** beautiful and well favoured."

Gen. 30:29 "And he said unto him, Thou knowest how I have served thee, and how thy cattle **was** with me."

At first I thought that this *HAYAH* might be expressing the location of the cattle. The cattle were with Jacob. But as I looked at the context, it became apparent that Jacob was not telling Laban about his cattle's location. There was more to it than that. Instead of reminding Laban of where his cattle had been, Jacob was reminding Laban of how his cattle had been. He was reminding Laban of how his cattle had prospered. He is emphasizing what **had become** of Laban's cattle. Let's see how other versions translate this *HAYAH*.

Gen. 30:29 "But he said to him, "You yourself know how I have served you and how your cattle have **fared** with me." (NASB)

Gen. 30:29 "Jacob said to him, "You yourself know how I have served you, and how your cattle have **fared** with me." (NRSV)

Gen. 30:29 "Jacob said to him, "You know how I have worked for you and how your livestock has **fared** under my care." (NIV)

Gen. 30:29 "And he said to him, Thou knowest how I have served thee, and what thy cattle has **become** with me." (DBY)

Once again we see *HAYAH* as a dynamic condition that is best translated as "become."

Gen. 31:40 *"Thus* I **was**; in the day the drought consumed me, and the frost by night; and my sleep departed from mine eyes."

Jacob is describing his physical sufferings during the days and nights of taking care of Laban's livestock. He is describing what had become of him during that time. He is describing the dynamics of his situation. He is saying, "this is what became of me. It wasn't "was" in the sense of this is what he always was.

Gen. 35:3 "And let us arise, and go up to Bethel; and I will make there an altar unto God, who answered me in the day of my distress, and **was** with me in the way which I went."

Gen. 35:5 "And they journeyed: and the terror of God **was** upon the cities that *were* round about them, and they did not pursue after the sons of Jacob."

Gen. 35:16 "And they journeyed from Bethel; and there **was** but a little way to come to Ephrath: and Rachel travailed, and she had hard labour."

Gen. 36:12 "And Timna **was** concubine to Eliphaz Esau's son; and she bare to Eliphaz Amalek: these *were* the sons of Adah Esau's wife."

Gen. 37:2 "These *are* the generations of Jacob. Joseph, *being* seventeen years old, **was** feeding the flock with his brethren; and the lad *was* with the sons of Bilhah, and with the sons of Zilpah, his father's wives: and Joseph brought unto his father their evil report."

Gen. 38:5 "And she yet again conceived, and bare a son; and called his name Shelah: and he **was** at Chezib, when she bare him."

Gen. 38:7 "And Er, Judah's firstborn, **was** wicked in the sight of the LORD; and the LORD slew him."

As the Hebrew considered goodness or uprightness, so they did with wickedness. It is a dynamic condition of the heart. If it were not so, no one could be saved.

Gen. 38:21 "Then he asked the men of that place, saying, Where *is* the harlot, that *was* openly by the way side? And they said, There **was** no harlot in this *place*."

Gen. 38:22 "And he returned to Judah, and said, I cannot find her; and also the men of the place said, *that* there **was** no harlot in this *place*."

Gen. 39:2 "And the LORD **was** with Joseph, and he **was** a prosperous man; and he **was** in the house of his master the Egyptian."

Gen. 39:5 "And it came to pass from the time *that* he had made him overseer in his house, and over all that he had, that the LORD blessed the Egyptian's house for Joseph's sake; and the blessing of the LORD **was** upon all that he had in the house, and in the field."

Gen. 39:6 "And he left all that he had in Joseph's hand; and he knew not ought he had, save the bread which he did eat. And Joseph **was** *a* goodly *person,* and well favoured."

Gen. 39:20 "And Joseph's master took him, and put him into the prison, a place where the king's prisoners *were* bound: and he **was** there in the prison."

Gen. 39:21 "But the LORD **was** with Joseph, and showed him mercy, and gave him favour in the sight of the keeper of the prison."

Gen. 39:22 "And the keeper of the prison committed to Joseph's hand all the prisoners that *were* in the prison; and whatsoever they did there, he **was** the doer *of it*."

Gen. 41:13 "And it came to pass, as he interpreted to us, so it **was**; me he restored unto mine office, and him he hanged."

This is a very important verse in helping us understand *HAYAH.* If you recall the situation in the Bible, Pharaoh had a troubling dream and wanted someone to interpret it. No one could. Then Pharaoh's butler remembered Joseph in prison and how he had correctly interpreted his and the baker's dreams. He tells Pharaoh that just as Joseph interpreted the dreams, so it **became**. What Joseph interpreted was what **came to be**. It wasn't "was" because it hadn't happened at the time Joseph interpreted the dreams. It wasn't until three days later that what he had interpreted came to pass. This *HAYAH* would better be translated "became" or "came to be."

Gen. 41:53 "And the seven years of plenteousness, that **was** in the land of Egypt, were ended."

Gen. 41:54 "And the seven years of dearth began to come, according as Joseph had said: and the dearth **was** in all lands; but in all the land of Egypt there **was** bread."

Gen. 41:56 "And the famine **was** over all the face of the earth: And Joseph opened all the storehouses, and sold unto the Egyptians; and the famine waxed sore in the land of Egypt."

Gen. 42:5 "And the sons of Israel came to buy *corn* among those that came: for the famine **was** in the land of Canaan."

Gen. 47:28 "And Jacob lived in the land of Egypt seventeen years: so the whole age of Jacob **was** an hundred forty and seven years."

Gen. 50:9 "And there went up with him both chariots and horsemen: and it **was** a very great company."

This *HAYAH* is used in connection with describing the size of the army that accompanied Joseph when he took Isaac's body back to the land of Caanan to be buried. As a way of describing the size of something, I can see why the ancient Hebrews would think in dynamic terms rather than static ones. The size of something is not equivalent to the thing itself. Armies aren't always the same size. Any particular army could increase or decrease in numbers over time, yet even with such dynamic change, it would still be an army.

WAST

Gen. 40:13 "Yet within three days shall Pharaoh lift up thine head, and restore thee unto thy place: and thou shalt deliver Pharaoh's cup into his hand, after the former manner when thou **wast** his butler."

HAYAH in Genesis 1:2

There you have it! We have just finished looking at the 68 instances of *HAYAH* in Genesis (other than Genesis 1:2) that are translated "was" or "wast" in the *King James Bible*. Over 30 of them express the idea of "becoming," "came," "came to pass," or "coming to be." If nearly one-half of the *HAYAH*s translated "was" in Genesis actually express the idea of "became," then how can anyone say that *HAYAH* cannot be translated "became?" Add these to the 139 times that *HAYAH* is already translated as some form of "became" and we see that *HAYAH* conveys "became" over 169 times out of its 300 or so uses. I think it makes "became" a viable translation.

Unfortunately, we aren't finished. We can't quit here. There is another aspect of *HAYAH* that we must study. We need to look at the form of the verb in Genesis 1:2 and compare how that same verb form is translated elsewhere. Since different forms of a verb can express different ideas, we must concentrate our study of the verb form in Genesis 1:2. Some argue that other forms of *HAYAH* may be translated "became" but that the verb form in Genesis 1:2 can't mean "became." We need to see if that is true. The *HAYAH* in Genesis 1:2 is in the *QAL* Perfect form of the verb. So, what does that mean? If you're like me, that doesn't mean much. Since I'm not a Hebrew scholar, we need to see what Hebrew scholars say.

QAL

"The Qal stem is the basic verbal stem in Hebrew language. Approximately two-thirds of the verbal forms in the Old Testament are in this stem. The Qal stem can be divided into two main classes: verbs that represent **action**... and verbs that describe **a state of being**..."[52]

There! Now we are Hebrew scholars! (But, I didn't say we were great Hebrew scholars.) We can say that the *QAL* is the most used verb form in the Old Testament, and that there is a division between *QAL* verbs of **action** and *QAL* verbs of **being**. Here is another clue to help us determine the meaning of *HAYAH* in Genesis 1:2. Is it in the form that represents ACTION or in the form that describes a STATE OF BEING? If it describes a state of being, then it is more likely that the earth **was** (in a state of being) without form, and void. If it describes action, then it is more likely that the earth **became** (came to be) without form, and void. As you can probably guess, it is in a form that describes action. There are two sub-groups of *QAL* verbs that represent action. These are the *QAL* Imperfect and the *QAL* Perfect. Let's look at what they mean.

QAL Imperfect

"The Qal Imperfect (qmf) indicates, in the active voice, simple imperfective **action**, viewed as part of a whole event or situation. 'If the priest that is anointed <u>do sin</u>...' (Lev. 4:3); 'And Moab <u>was sore afraid</u> of the people...' (Num. 22:3)."[53]

The *QAL* Imperfect form of a verb describes an action. An imperfect verb tends to describe action as continuous or incomplete with respect to time. The *QAL* Imperfect of *HAYAH* would describe an **action** with respect to time rather than a **being** with respect to time. Therefore, *HAYAH* in the *QAL* Imperfect should not be translated "was" in the sense of a state of being. In fact, that's exactly what we see when we look at its translations. We will first look at the *QAL* Imperfect in its simple form and then we will look at a sub-category of the *QAL* Imperfect. There are over forty uses of the simple form of the *QAL* Imperfect of *HAYAH* in Genesis, but only one of them is translated "was" in the *King James Bible*. All the others are translated "shall be," "will be," or some variation of a "yet-to-be" ongoing action. Genesis 44:17 is a good example of how the *QAL* Imperfect of *HAYAH* is usually translated.

Gen. 44:17 "But he said, 'Far be it from me that I should do so! Only the man in whose hand the cup was found **shall be** my servant. But as for you, go up in peace to hour father." (KJV)

If you remember the story, Joseph set his brothers up for a test. He had his own royal cup placed into Benjamin's sack of grain before they set out from Egypt to return to their father. Then when the cup was found, it would look as if Benjamin had stolen it. Joseph was telling his brothers that the one who had "stolen" the cup would BECOME his slave. The *QAL* Imperfect of *HAYAH* is used because it describes what Benjamin was to BECOME, not what Benjamin WAS. Again, the *QAL* Imperfect of *HAYAH* should not be translated as a static condition of being.

Now, you may be wondering why I am mentioning the *QAL* Imperfect of *HAYAH* when the *HAYAH* in Genesis 1:2 is in the *QAL* Perfect. There are two reasons. First, I want you to see how the Hebrews thought when they used *HAYAH*. Even though the form of the verb is different, the verb itself has the same basic underlying concept. That concept was of a dynamic being or condition. Secondly, there is a verse in Genesis where the *King James* translators translated the *QAL* Imperfect of *HAYAH* as "was," and I don't want to skip a "was" in Genesis. I don't want you to think that I'm picking and choosing just the verses I want you to see. I don't want you to think I am misleading you. Genesis 2:5 is the only place in Genesis where the *QAL* Imperfect of *HAYAH* is translated "was" in the *King James Version*. We need to see if this "was" expresses a dynamic or a static state of being.

Gen. 2:5 "And every plant of the field before it **was** in the earth, and every herb of the field before it grew: for the LORD God had not caused it to rain upon the earth, and *there was* not a man to till the ground." (KJV)

We've already looked at this verse and saw that it describes an action. It is describing a dynamic condition. This verse describes the earth before plants "came into being." Look at how the *New International Version* translates it.

Gen. 2:5 "and no shrub of the field **had yet appeared** on the earth and no plant of the field had yet sprung up, for the LORD God had not sent rain on the earth and there was no man to work the ground," (NIV)

HAYAH is a verb of action here. No shrub had yet appeared; no plant had yet sprung up. As I mentioned before, God describes the earth without plants as a dynamic condition, but He describes the earth without man as a static condition. No *HAYAH* is used in the phrase about man. Can you begin to get a feel for God's view of earth's condition before man's creation? Yes, these are God's words. He picked the way the Pre-Adamic earth should be described. He describes earth without man in a different way than He describes earth without plants. Earth had always been without man, but earth **came to be** without plants. One condition was static; the other condition was dynamic. Are we willing to listen to God?

So that you won't feel cheated, I am going to list all the *QAL* Imperfect uses of *HAYAH* in Genesis and show how they are translated in the *King James Version*. Again, I'm doing this so you can get a better feel for what went through the minds of the ancient Hebrews.

QAL Imperfect of *HAYAH* in Genesis (KJV)

be [2] 9:11, 38:23
become [1] 9:15
keep [1] 33:9
may be [1] 21:30
shall be [25] 1:29, 6:19, 9:2, 9:3, 9:25, 15:5, 15:13, 17:16, 27:39, 28:22. 31:8. 31:8, 34:10, 35:10, 35:11, 41:27, 41:36, 44:10, 44:10, 44:17, 47:24, 48:5, 48:6, 49:17, 49:26
shall become [2] 48:19, 48:19
shalt be [2] 4:12, 41:40
should be [1] 38:9
will be [4] 16:12, 28:20, 34:15, 44:9
will become [1] 37:20
was [1] 2:5

Genesis 2:5 does not convey the idea of static being even though it is translated "was." Now, this isn't the only place where *HAYAH* in the *QAL* Imperfect is translated "was." There are other instances of the *QAL* Imperfect of *HAYAH* that are translated "was," but they are not in the same form; they belong to a sub-category of the *QAL* Imperfect. This sub-category is the *WAW* Consecutive. What is a *WAW* Consecutive?

WAW Consecutive

"The *Waw* Consecutive (wcs) if two verbs are referring to the past in one continuous narration, only the first verb is in the Perfect, while any following verb is in the Imperfect with a prefixed *waw*. 'And Judah said unto Simeon... I likewise will go with thee into the lot. So Simeon went with him' (Judg. 1:3); '... Yet ye say, Wherein hast thou loved us? ... yet I love Jacob' (Mal. 1:2). Conversely, in a continuous narration referring to the future only the first verb is in the Imperfect, while any following verb is in the Perfect with a prefixed *waw*. '... Let there be sought for my lord the king a young virgin: and let her stand before the king...' (1 Kgs. 1:2); '... bring *it* into me, and I will hear it" (Deut. 1:7)." [54]

Below are the instances of *HAYAH* in the *WAW* Consecutive, *QAL* Imperfect in Genesis that are translated "was" in the *King James Bible*.

Gen. 1:3	Gen. 1:7	Gen. 1:9	Gen. 1:11
Gen. 1:15	Gen. 1:24	Gen. 1:30	Gen. 4:2
Gen. 5:32	Gen. 7:12	Gen. 7:17	Gen. 10:10
Gen. 10:19	Gen. 10:30	Gen. 11:1	Gen. 11:30
Gen. 12:10	Gen. 13:7	Gen. 17:1	Gen. 21:20
Gen. 23:1	Gen. 25:20	Gen. 25:27	Gen. 26:1

Gen. 26:34	Gen. 27:30	Gen. 35:3	Gen. 35:5
Gen. 35:16	Gen. 38:7	Gen. 39:2	Gen. 39:2
Gen. 39:2	Gen. 39:5	Gen. 39:6	Gen. 39:20
Gen. 39:21	Gen. 41:54	Gen. 47:28	Gen. 50:9

We have looked at these already, but you should review them. Since they are in a form that represents action rather than being, we need to interpret them as conditions of action and not conditions of being. Again, I know that Genesis 1:2 doesn't use the *QAL* Imperfect form of *HAYAH*; it uses the *QAL* Perfect form, but I want to spend some time on this. I want you to see how the ancient Hebrews used this word. I want you to begin to get a glimpse of what they thought about *HAYAH*. I want you to think like an ancient Hebrew.

Since the *HAYAH* in Gen. 1:2 is in the *QAL* Perfect form, we need to look at what the *QAL* Perfect form means.

QAL Perfect

"The Qal Perfect (qpf) indicates, in the active voice, simple perfective **action**, viewed as a whole. 'I will sing unto the Lord, because he hath dealt bountifully with me.' (Ps. 13:6); 'For a nation is come up upon the land…' (Joel 1:6)."[55] (bold emphasis mine)

Again we see a form of the *QAL* stem that describes **ACTION RATHER THAN BEING.** In this case, the action is usually viewed as completed. This is the form found in Genesis 1:2. Before we look at the *QAL* Perfect, let's examine a sub-category of the *QAL* Perfect, the *WAW* Conjunctive of the *QAL* Perfect. (Are you confused yet?) Let's find out what the *WAW* Conjunctive means.

WAW Conjunctive

"The *Waw* Conjunctive (wcj) The letter *waw* serves as a link between two words, clauses, or sentences, and is affixed inseparably to the word that follows it. It is normally pointed with sh^ewa, but may take other vowels depending on which letter of the alphabet it precedes. '... for Joseph was in Egypt *already*' (Ex 1:5); '... the nobles and princes of the provinces...' (Esth. 1:3). Its meaning with verbs is so decisive that some view it as a separate conjunction. '... Be fruitful, and multiply, and fill the waters in the seas...' (Gen. 1:22); 'Behold ye among the heathen, and regard, and wonder marvelously...' (Hab. 1:5)"[56]

WAW Conjunctive of the *QAL* Perfect of *HAYAH* in Genesis (KJV)

came to pass [1] 30:41   **come to pass** [1] 24:14

shall be [2] 9:13, 48:21   **shall come to pass** [5] 9:14, 24:43, 44:31, 46:33, 47:24

was [1] 38:5

The only reason I mention the *WAW* Conjunctive of the *QAL* Perfect is because the *King James* translators translated one such *HAYAH* in Genesis as "was," and I want to thoroughly examine all the "was" translations of *HAYAH* in Genesis. I don't want to leave even one out. All the other uses of *HAYAH* in the *WAW* Conjunctive of the *QAL* Perfect are translated something else. There is one verse in the *King James Version* of Genesis that translates the *WAW* Conjunctive, *QAL* Perfect form of *HAYAH* as "was." That verse is Genesis 38:5.

Gen. 38:5 "And she yet again conceived, and bare a son; and called his name Shelah: and he **was** at Chezib, when she bare him."

As we have seen, it describes the dynamic condition of location. Here again, even though the *HAYAH* of Genesis 38:5 is translated "was," it does not mean a static condition, being, or existence. It is dynamic.

The *HAYAH* of Genesis 1:2 (The *QAL* Perfect)

Finally we get to the *HAYAH* of Genesis 1:2. It is in the *QAL* Perfect form. This means that generally, it should be translated as a form of "to be" that implies a past completed action and not a past static condition. Again, both the Imperfect and Perfect forms of the *QAL* stem imply ACTION rather than BEING. I will list all the "was" translations of *HAYAH* in Genesis that are in the *QAL* Perfect form. Following each verse in the KJV (*King James Version*) I will list the PSV (*Peasant Steve Version*) (underlined) to show how a form of "became" makes better sense. I'll save Genesis 1:2 for last.

Gen. 3:1 "Now the serpent **was** more subtle than any beast of the field which the LORD God had made. And he said unto the woman, Yea, hath God said, Ye shall not eat of every tree of the garden?"

<u>Gen. 3:1 "Now the serpent **came to be** more subtle than any beast of the field which the LORD God had made. And he said unto the woman, Yea, hath God said, Ye shall not eat of every tree of the garden?"</u>

Gen. 3:20 "And Adam called his wife's name Eve; because she **was** the mother of all living."

Gen. 3:20 "And Adam called his wife's name Eve; because she **became** the mother of all living."

Gen. 4:2 "And she again bare his brother Abel. And Abel was a keeper of sheep, but Cain **was** a tiller of the ground."

Gen. 4:2 "And she again bare his brother Abel. And Abel was a keeper of sheep, but Cain **became** a tiller of the ground."

Gen. 4:20 "And Adah bare Jabal: he **was** the father of such as dwell in tents, and *of such as have* cattle."

Gen. 4:20 "And Adah bare Jabal: he **became** the father of such as dwell in tents, and *of such as have* cattle."

Gen. 4:21 "And his brother's name *was* Jubal: he **was** the father of all such as handle the harp and organ."

Gen. 4:21 "And his brother's name *was* Jubal: he **became** the father of all such as handle the harp and organ."

Gen. 6:9 "These *are* the generations of Noah: Noah **was** a just man *and* perfect in his generations, *and* Noah walked with God."

Gen. 6:9 "These *are* the generations of Noah: Noah **became** a just man *and* perfect in his generations, *and* Noah walked with God."

Gen. 7:6 "And Noah *was* six hundred years old when the flood of waters **was** upon the earth."

Gen. 7:6 "And Noah *was* six hundred years old when the flood of waters **came** upon the earth."

Gen. 10:9 "He **was** a mighty hunter before the LORD: wherefore it is said, Even as Nimrod the mighty hunter before the
LORD."

<u>Gen. 10:9 "He **became** a mighty hunter before the LORD:
wherefore it is said, Even as Nimrod the mighty hunter
before the LORD."</u>

Gen. 13:6 "And the land was not able to bear them, that they
might dwell together: for their substance **was** great, so that
they could not dwell together."

<u>Gen. 13:6 "And the land was not able to bear them, that they
might dwell together: for their substance **became** great, so
that they could not dwell together."</u>

Gen. 15:17 "And it came to pass, that, when the sun went
down, and it **was** dark, behold a smoking furnace, and a
burning lamp that passed between those pieces."

<u>Gen. 15:17 "And it came to pass, that, when the sun went
down, and it **became** dark, behold a smoking furnace, and a
burning lamp that passed between those pieces."</u>

Gen. 26:1 "And there was a famine in the land, beside the
first famine that **was** in the days of Abraham. And Isaac went
unto Abimelech king of the Philistines unto Gerar."

<u>Gen. 26:1 "And there was a famine in the land, beside the
first famine that **came** in the days of Abraham. And Isaac
went unto Abimelech king of the Philistines unto Gerar."</u>

Gen. 26:28 "And they said, We saw certainly that the LORD
was with thee: and we said, Let there be now an oath betwixt

us, *even* betwixt us and thee, and let us make a covenant with thee;"

Gen. 26:28 <u>"And they said, We saw certainly that the LORD **came to be** with thee: and we said, Let there be now an oath betwixt us, *even* betwixt us and thee, and let us make a covenant with thee;"</u>

Gen. 29:17 "Leah *was* tender eyed; but Rachel **was** beautiful and well favoured."

Gen. 29:17 <u>"Leah *was* tender eyed; but Rachel **came to be** beautiful and well favoured."</u>

Gen. 30:29 "And he said unto him, Thou knowest how I have served thee, and how thy cattle **was** with me."

Gen. 30:29 <u>"And he said unto him, Thou knowest how I have served thee, and how thy cattle **came to be** with me."</u>

Gen. 31:40 *"Thus* I **was**; in the day the drought consumed me, and the frost by night; and my sleep departed from mine eyes."

Gen. 31:40 <u>*"Thus* I **became**; in the day the drought consumed me, and the frost by night; and my sleep departed from mine eyes."</u>

Gen. 36:12 "And Timna **was** concubine to Eliphaz Esau's son; and she bare to Eliphaz Amalek: these *were* the sons of Adah Esau's wife."

Gen. 36:12 <u>"And Timna **became** concubine to Eliphaz Esau's son; and she bare to Eliphaz Amalek: these *were* the sons of Adah Esau's wife."</u>

Gen. 37:2 "These *are* the generations of Jacob. Joseph, *being* seventeen years old, **was** feeding the flock with his brethren; and the lad *was* with the sons of Bilhah, and with the sons of Zilpah, his father's wives: and Joseph brought unto his father their evil report."

Gen. 37:2 "These *are* the generations of Jacob. Joseph, *being* seventeen years old, **came** feeding the flock with his brethren; and the lad *was* with the sons of Bilhah, and with the sons of Zilpah, his father's wives: and Joseph brought unto his father their evil report."

Gen. 38:21 "Then he asked the men of that place, saying, Where *is* the harlot, that *was* openly by the way side? And they said, There **was** no harlot in this *place*."

Gen. 38:21 "Then he asked the men of that place, saying, Where *is* the harlot, that *was* openly by the way side? And they said, There **came** no harlot in this *place*."

Gen. 38:22 "And he returned to Judah, and said, I cannot find her; and also the men of the place said, *that* there **was** no harlot in this *place*."

Gen. 38:22 "And he returned to Judah, and said, I cannot find her; and also the men of the place said, *that* there **came** no harlot in this *place*."

Gen. 39:22 "And the keeper of the prison committed to Joseph's hand all the prisoners that *were* in the prison; and whatsoever they did there, he **was** the doer *of it*."

Gen. 39:22 "And the keeper of the prison committed to Joseph's hand all the prisoners that *were* in the prison; and whatsoever they did there, he **became** the doer *of it*."

Gen. 41:13 "And it came to pass, as he interpreted to us, so it **was**; me he restored unto mine office, and him he hanged."

Gen. 41:13 "And it came to pass, as he interpreted to us, so it **became**; me he restored unto mine office, and him he hanged."

Gen. 41:53 "And the seven years of plenteousness, that **was** in the land of Egypt, were ended."

Gen. 41:53 "And the seven years of plenteousness, that **came** in the land of Egypt, were ended."

Gen. 41:54 "And the seven years of dearth began to come, according as Joseph had said: and the dearth was in all lands; but in all the land of Egypt there **was** bread."

Gen. 41:54 "And the seven years of dearth began to come, according as Joseph had said: and the dearth was in all lands; but in all the land of Egypt there **came to be** bread."

Gen. 41:56 "And the famine **was** over all the face of the earth: And Joseph opened all the storehouses, and sold unto the Egyptians; and the famine waxed sore in the land of Egypt."

Gen. 41:56 "And the famine **came** over all the face of the earth: And Joseph opened all the storehouses, and sold unto the Egyptians; and the famine waxed sore in the land of Egypt."

Gen. 42:5 "And the sons of Israel came to buy *corn* among those that came: for the famine **was** in the land of Canaan."

Gen. 42:5 "And the sons of Israel came to buy *corn* among those that came: for the famine **came** in the land of Canaan."

Here we see the twenty-five instances of *HAYAH* in the *QAL* Perfect form that are translated "was" in Genesis in the *King James Bible*. Again, this is the same form as in Genesis 1:2. I can substitute "became," "came," "came to be," or some other form of "become" in all of them and they make as much sense, or even more sense than the word "was." In other words, every instance of *HAYAH* in the *QAL* Perfect form that is translated "was" in Genesis in the *King James Bible* describes a dynamic "becoming" rather than a static "being." Someone tell me again that the *QAL* Perfect form of *HAYAH* CANNOT be translated as a dynamic condition of becoming.

By now your brain might be a little overworked by all this talk about Hebrew rules of grammar. I know it is confusing and tedious, and you might have been tempted to skim over it too quickly. Please, stop for a moment and read this: **In The Book of Genesis, every instance of *HAYAH* in the *QAL* Perfect form (the same form that is used in Genesis 1:2) that is translated "was" in the *King James Bible*, actually describes a dynamic condition of becoming and not a static condition of being.** So, what is God telling us in Genesis 1:2? What is the best INFORMATIONAL translation? Only the Gap Theory interprets Genesis with the information that God revealed.

Gen. 1:2 "But the earth **had become** without form, and void; and darkness was upon the face of the deep. And the Spirit of God moved upon the face of the waters." (PSV)

Summary of the Evidence for the Restoration Theory

1. Qualified Hebrew scholars confirm that *HAYAH* can be translated "became." The Restoration Theory is not based on an unscholarly interpretation.
2. Job 38:4-6 describes the creation of the earth as a process over time. The Restoration Theory agrees.

3. The Restoration Theory has been believed and defended by theologians for hundreds, if not thousands of years. It is not something dreamed up to compromise the Bible with evolution.

4. Isaiah 45:18 reveals that God did not create the earth a desolate waste. This is what the Restoration Theory teaches.

5. Job 38:7 shows that the angels rejoiced over the creation of the earth, thus indicating that the earth wasn't created a desolate waste. The Restoration Theory agrees with this.

6. Job 38:8-11 reveals that there was a global flood accompanied by global darkness that happened before Adam was created. This is one of the beliefs of the Gap Theory.

7. Paul indicates in 2 Corinthians 4:6 that God had already created light in the universe before Genesis 1:3. Light was created at the first beginning. The Gap Theory is in agreement.

8. According to the Bible, the creation of the earth came before the first day of creation. Genesis 1:1-2 is not a title or summary of creation. This is at the heart of the Restoration Theory.

9. Jeremiah 4:23-26 reveals that the Pre-Adamic earth was made *TOHUW WAW BOHUW* by God's fierce anger. The Restoration Theory says that the earth was judged by God's fierce anger.

10. Isaiah 34:11 confirms that *TOHUW* and *BOHUW* describe Divine judgment. The Restoration Theory agrees again.

11. Interpreting *TOHUW WAW BOHUW* to mean something other than Divine judgment violates accepted rules of letting Scripture interpret Scripture. The Restoration Theory doesn't violate this rule. Rather, it enforces it.

12. The Restoration Theory explains why God pronounced that His works of restoration were "good," but did not say that of the heavens and the earth in Genesis 1:2.

13. The writings of Job and Ezekiel combined with the teachings of Jesus and John show that earth had two separate beginnings. The Restoration Theory prevents a contradiction between what Jesus and John taught about Lucifer and what Job and Ezekiel taught.

14. The descriptions of what Lucifer did to the earth match with the descriptions of what God did to the Pre-Adamic earth as a result of sin. This fits with the Restoration Theory.

15. Both Haggai 2:6 and Hebrews 12:25-27 reveal that God has previously judged (shaken) the Pre-Adamic heavens and earth. Only the Restoration Theory mentions a Pre-Adamic judgment.

16. The Restoration Theory explains the discrepancy between the order of fossils seen in the Geological Column and the order of creation/restoration mentioned in Genesis 1:3-31. God did not restore life in the same order He had originally created it.

17. The Restoration Theory explains the phenomenon of "Living Fossils." Some things had died millions of years ago and God restored them recently.

18. The Restoration Theory explains why some dating techniques indicate a Young-Earth while other dating techniques indicate an Old-Earth.

19. The Gap Theory actually explains the "gaps" in the fossil record. Geological evidence repeatedly shows both the sudden appearance and the sudden disappearance of species. The Bible gives no clues as to when or how often God created Pre-Adamic life forms. He may have periodically visited earth to create new species. At the same time, Lucifer and his angels may have been responsible for the sudden kill off of old species.

20. The meaning of *YOWM* in Genesis One can only mean a literal twenty-four hour day. The Restoration Theory

and the Young-Earth Theory agree with the Bible on this meaning, but the Day-Age Theory doesn't.

21. The observations of science indicate that the universe is old. The Restoration Theory and the Day-Age Theory agree with these scientific observations, but the Young-Earth Theory doesn't.

22. The Restoration Theory reveals there were intelligent beings on the earth before man. This is what paleontologists seem to have discovered. Science doesn't disagree with Scripture.

23. The Ruin-Restoration Theory teaches that the judgment and restoration of the earth was a foreshadow of what Christ does for fallen man. The visible heavens declare the invisible glory of our Redeemer. Redemption is not seen in the heavens if they were never redeemed.

24. Jeremiah saw the earth in a condition of having dry land but no light while Genesis shows the earth in a condition of having light but no dry land. There is no contradiction here; the Gap Theory explains the proper historical sequence of events.

25. Hebrews 11:3 reveals that God created an age before our present age. This was the Pre-Adamic age that the Gap Theory presents.

26. The New Testament mentions the "catabolism of the world." This was what God did when He made the earth *TOHUW WAW BOHUW* as explained by the Ruin-Restoration Theory.

So now I think you see my motive for believing in the Restoration Theory. I think you understand **why** I believe what I believe; not just **what** I believe. It's not a matter of being able to manipulate Hebrew grammar rules. It is not a matter of trying to make the Bible agree with The Theory of Evolution. It doesn't! It is a matter of seeing how Scripture agrees with Scripture. It is a matter of showing how God's Works agree with God's Words. It is a matter of giving Jesus Christ all the Glory He so richly deserves. He is the God of Mercy, Grace, Forgiveness, and most of all, the God of Restoration

It is amazing how the Gap Theory is rejected as a weak argument by Young-Earth and Day-Age creationists. Their theories have far less evidence. Their theories create problems and confusion. If you study the other creation theories with an open heart, I think you will discover that they have little biblical and scientific evidence in their favor. Why then does the Restoration Theory receive such criticism? Satan doesn't want us to see how the handiwork of God reveals His invisible attribute of Redeemer/Restorer. The Gap Theory glorifies God by revealing how His restoration of the fallen earth foreshadows His restoration of fallen man. **Unbelievers, please listen to me. Believe on the Lord Jesus Christ, and you will be restored.**

Dear unsaved friends: I know some of you have rejected the Bible because you believe it is full of lies and errors. I understand your rejection. I wouldn't believe it either if it taught things that weren't true. If you have read this book with an open mind, then I think you now realize that the Bible doesn't teach many of the things that you have been told, things you objected to, things you knew weren't true! The Bible reveals the truth. The Bible reveals the truth about the origin of man, and more importantly, the Bible reveals the truth about the destiny of man. Most importantly, the Bible reveals the truth about YOUR destiny. God will not tolerate sin. He will not tolerate your sin and He will not tolerate my sin. We all have sinned, and we are doomed to face God's eternal judgment if we reject God's gift of forgiveness and salvation through Jesus Christ. All of us are fallen beings in need of restoration, and that restoration comes only from the God who restores. That restoration comes only by faith in Jesus Christ.

Here is the most important truth you need to know: God loved you enough to become your substitute for His own wrath against your sin. He came as a human baby born in the most humble of circumstances. He became a Dirt-Man. He lived a human life full of the same suffering, sorrow, pain, and temptations that we all face. However, He did it without succumbing to sin. He lived a perfect life, but He died a most horrible death. When Jesus went to the cross, it wasn't a symbolic act of self-sacrifice. It was to pay in full the price for our sins. When He screamed out, "My God, My God, Why have you forsaken Me," it wasn't due to the physical pain of the nails in His hands and feet. He cried out because God was judging Him in our place. For three hours the Father forsook His beloved Son. For three hours God caused darkness to cover the land. He caused the same kind of darkness that covered the pre-Adamic earth. In those three hours of darkness, Jesus was judged in our place and bore an eternity of hell for each and every one of us. He did this because He didn't want you to have to bear it. He didn't want you to suffer eternal death. The Bible is very clear about how we must be saved. Believe that Jesus went to the cross to be your substitute for the punishment for your sins, and you will be saved.

John 3:16 "For God so loved the world, that He gave His only begotten Son, that whoever believes in Him should not perish, but have eternal life." (NASB)

General References

—⚇—

1. Dill, Steven E.: *Objections to the Doctrine of Evolution*
   Copyright 1995: Steven E. Dill; Louisville, Kentucky
   Unpublished at this time

2. Kimball, John W.: *Biology* 2nd Edition
   Copyright: Addison-Wesley Publishing Co. 1969

3. Carpenter, Kenneth: *Eggs, Nests, and Baby Dinosaurs*
   Copyright Kenneth Carpenter 1999
   Bloomington, Indiana: Indiana University Press

4. Hayward, Alan: *Creation and Evolution*
   Copyright: Alan Hayward. 1985
   Minneapolis: Bethany House Publishers

5. Quoted by Arthur C. Custance, *Without Form and Void*
   Copyright: Arthur C. Custance, 1970
   Brockville, Ontario: Doorway Publications, page 130

6. Quoted by Arthur C. Custance, *Without Form and Void*
   Copyright: Arthur C. Custance, 1970
   Brockville, Ontario: Doorway Publications, page 130

7. Quoted by Arthur C. Custance, *Without Form and Void*
   Copyright: Arthur C. Custance, 1970
   Brockville, Ontario: Doorway Publications, page 131

8. Dillman, August, *Genesis Critically and Exegetically Expounded*
(Translation by William B. Stevenson)
Edinburgh: T. & T. Clark, 1897

9. Young, Robert, *Literal Translation of the Bible*,
Public Domain

10. Thieme, Robert B., Jr., *Creation, Chaos, and Restoration*
Copyright: Robert B. Thieme, Jr., 1974
Houston: Berachah Tapes and Publications, page 23

11. Quoted by Arthur C. Custance, *Without Form and Void*
Copyright: Arthur C. Custance, 1970
Brockville, Ontario: Doorway Publications, page 131

12. Quoted by Arthur C. Custance, *Without Form and Void*
Copyright: Arthur C. Custance, 1970
Brockville, Ontario: Doorway Publications, page 131

13. Custance, Arthur C., *Without Form and Void*
Copyright: Arthur C. Custance, 1970
Brockville, Ontario: Doorway Publications, page xi

14. Quoted by Arthur C. Custance, *Without Form and Void*
Copyright: Arthur C. Custance, 1970
Brockville, Ontario: Doorway Publications, page 131

15. Quoted by Arthur C. Custance, *Without Form and Void*
Copyright: Arthur C. Custance, 1970
Brockville, Ontario: Doorway Publications, page 132

16. Strong, James, *Strong's Dictionary of the Hebrew Language*
Copyright: James Strong, 1890
London: Hodder and Stroughton

17. *The New American Standard Exhaustive Concordance of the Bible*
Copyright 1981: The Lockman Foundation

18. Strong, James, *Strong's Dictionary of the Hebrew Language*
Copyright: James Strong, 1890
London: Hodder and Stroughton

19. *The Complete Word Study Old Testament King James Version*
Copyright: 1994 AMG International, Inc. D/B/A AMG Publishers
Chattanooga, TN 37422, U.S.A.

20. Scofield, C. I., *The New Scofield Reference Bible*
New York London Toronto: Oxford University Press, Inc., 1967

21. Larkin, Clarence, *Dispensational Truth*
Copyright: Clarence Larkin, 1920
Philadelphia: Rev. Clarence Larkin Est.

22. Thieme, Robert B., Jr., *Creation, Chaos, and Restoration*
Copyright: Robert B. Thieme, Jr., 1974
Houston: Berachah Tapes and Publications, pages 1-12

23. Dake, Finis Jennings, *Dake's Annotated Reference Bible*
Copyright: Finis Jennings Dake, 1963
Lawrenceville, Georgia: Dake Bible Sales, Inc.

24. Sauer, Eric, (1931-) *The King of the Earth*
London: Paternoster Press, 1970

25. Pember, G. H. Pember, (1837-1910) *Earth's Earliest Ages*
London: Hodder and Stroughton, 1876
Grand Rapids: Kregel Publications

26. Ramm, Bernard Ramm, *The Christian View of Science and Scripture*
London: Paternoster Press, 1964

27. Darby, John Nelson, (1800-1882) *The Collected Writings in 34 Volumes*, Edited by William Kelly
Public Domain

28. Jamieson, Robert; Fausset, A. R.; Brown, David, *Jamieson, Fausset, Brown Commentary Critical and Explanatory on the Whole Bible*,
Public Domain, 1871

29. Edersheim, Alfred, *Bible History, Old Testament*, (1890)
Public Domain
Book 1: The World Before the Flood, and the History of the Patriarchs.
Hendrickson Publishers, Inc. 1995

30. Pink, A. W., *Gleanings in Genesis*, 1922
Public Domain,

31. Chafer, Lewis Sperry, *Systematic Theology, Volume 2*, p. 39
Copyright: Lewis Sperry Chafer, 1947
Dallas: Dallas Seminary Press, Sixth Printing, December 1969

32. Archer, Gleason, *A Survey of Old Testament Introduction*
Copyright: Gleason Archer, 1974
Chicago: Moody Press

33. Anstey, Martin, *The Romance of Bible Chronology*, London, 1913, p. 62
Marshal Brothers, LTD., London, Edinburgh, and New York, 1913

34. Warren Baker and Eugene Carpenter, *The Complete Word Study Dictionary: Old Testament*, p. 262
Copyright: 2003 AMG Publishers
Chattanooga, TN 37421

35. Coates, Charles Andrew, (1862-1945)
An Outline of the Book of Genesis
Kingston-on-Thames, Surrey: Stow Hill Bible and Tract Depot, 1921.

36. Kelly, William, (1821-1906)
Lectures on the Pentateuch
London: W. H. Broom, 1871 (Public Domain)

37. Gaebelein, Arno Clement, (1861-1945)
The Annotated Bible, the Book of Genesis
Copyright 1919 Arno Clement Gaebelein
(Public Domain)

38. Grant, Frederick William, (1834-1902)
Genesis in the Light of the New Testament
New York: Loizeaux Brothers. (Public Domain)

39. Hole, Frank Binford (1874-1964)
Genesis (Public Domain)

40. Barnes, Albert, (1798-1870)
Notes on the Bible (Public Domain)

41. Ginzberg, Louis, (1873-1953)
The Legends of The Jews, The Creation of the World, p. 4
Copyright 1912: The Jewish Publication Society,
Philadelphia

42. Custance, Arthur C., *Without Form and Void*
Copyright: Arthur C. Custance, 1970
Brockville, Ontario: Doorway Publications

43. Custance, Arthur C., *Without Form and Void*, p. 102
Copyright: Arthur C. Custance, 1970
Brockville, Ontario: Doorway Publications

44. Han, Nathan E., *A Parsing Guide to the Greek New Testament*, (Sixth Printing)
Copyright: Herald Press, 1971
Scottsdale, Pennsylvania

45. Custance, Arthur C., *Without Form and Void*, p. 18 and
Appendix XIX
Copyright: Arthur C. Custance, 1970
Brockville, Ontario: Doorway Publications

46. *The Wollemi Pine Tree*
Copyright: The National Geographic Society, 2006
Washington, D.C.

47. M. Krings, A. Stone, R. W. Schmitz, H. Krainitzki, M.
Stoneking, and S. Pääbo, July 1997.
Neanderthal DNA Sequences and the Origin of Modern
Humans
Cell 90:19-30.

48. Igor Ovchinnikov, A. Götherström, G. P. Romanoval, V. M. Kharitonov, K. Lidén, and W. Goodwin, March 2000.
Molecular Analysis of Neanderthal DNA from the Northern Caucasus
Nature 404:490-493.

49. James Strong, *Strong's Dictionary of the Hebrew Language*
Copyright: James Strong, 1890
London: Hodder and Stroughton

50. Anstey, Martin, *The Romance of Bible Chronology*, London, 1913, p. 62
Marshal Brothers, LTD., London, Edinburgh, and New York, 1913

51. *The Complete Word Study Old Testament King James Version*, p. 2311
Copyright : 1994 AMG International, Inc. D/B/A AMG Publishers
Chattanooga, TN 37422, U.S.A.

52. *The Complete Word Study Old Testament King James Version*, p. 2282
Copyright: 1994 AMG International, Inc. D/B/A AMG Publishers
Chattanooga, TN 37422, U.S.A.

53. *The Complete Word Study Old Testament King James Version*, p. 2282
Copyright : 1994 AMG International, Inc. D/B/A AMG Publishers
Chattanooga, TN 37422, U.S.A.

54. *The Complete Word Study Old Testament King James Version*, p. 2283
    Copyright: 1994 AMG International, Inc. D/B/A AMG Publishers
    Chattanooga, TN 37422, U.S.A.

55. *The Complete Word Study Old Testament King James Version*, p. 2283
    Copyright: 1994 AMG International, Inc. D/B/A AMG Publishers
    Chattanooga, TN 37422, U.S.A.

56. *The Complete Word Study Old Testament King James Version*, p. 2283
    Copyright: 1994 AMG International, Inc. D/B/A AMG Publishers
    Chattanooga, TN 37422, U.S.A.

Index

—m—

Custance, Arthur: 127, 138, 142, 187, 188, 234, 235, 305, 306, 308, 312
Dake, Finis Jennings: 178
Darby, John Nelson: 179
Day-Age Solutions: 63
Dillman, August: 126, 138, 143
Earth's Two Beginnings: 244
Edersheim, Alfred: 179
Edgar, King of England: 178
Episcopius, Simon: 178
Flood-Geology: 34, 80, 100, 101, 102, 105, 113, 122
Fossil Index System: 57, 58, 60. 61. 111, 113
Functional Age: 91
Gaebelein, Arno Clement: 185
Ginsberg, Louis: 186
Grant, Frederick William: 185
Green River Varves: 106
HAYAH (to be): 138, 139, 140, 163, 164, 174, 175, 181, 182, 183, 184, 186, 187, 188, 189, 212, 234, 235, 290, 291, 332, 333, 334, 335, 336, 337, 338, 339, 340, 341, 342, 343, 344, 345, 346, 348, 349, 350, 351, 353, 354, 355, 356, 359, 360, 361, 362, 363, 364, 365, 366, 367, 368, 374
Hayward, Alan: 122
Hole, Frank Binford: 185
Jahn, Herb: 138
Jamieson, Fausset, and Brown: 179
Japanese Plesiosaur: 116
Johanson, Donald: 331
KATABOLE KOSMOS: 306, 307, 309, 310, 316
KATABOLE SPERMA: 309, 311, 312, 314, 316
KATARTIDZO: 296, 297
Kelly, William: 184, 384
Large Magellanic Cloud: 93, 94
Larkin, Clarence: 175

Supernova: 93, 94
Thieme, Robert B.: 126, 133, 138, 141, 143, 178
TOHUW (without form): 144, 145, 154, 156, 157, 161, 177,
    191, 192, 193, 199, 200, 227, 228, 232, 233, 235, 237,
    238, 239, 243, 245, 266, 269, 326, 327
Traditional-Geology: 100, 101, 113
Two Global Floods: 290, 291
Ussher, James: 81, 82
Von Bohlem, Peter: 125, 132, 142
WAW (and, but): 134, 135, 136, 144, 161, 177, 183, 227,
    228, 232, 233, 235, 237, 238, 239, 245, 266, 269, 326,
    327, 365, 366, 367
YATSAR (form): 160, 199, 202, 326
Yellowstone National Park: 100, 101
Young, Robert: 126, 132, 138, 143
Young-Earth Solutions: 80

CPSIA information can be obtained
at www.ICGtesting.com
Printed in the USA
LVHW01s0044091117
555586LV00001B/75/P

9 781609 571634